GRADES 7-8

the Super Source™
Probability and Statistics

Cuisenaire Company of America, Inc.
White Plains, New York

Cuisenaire extends its warmest thanks to the many teachers and students across the country who helped ensure the success of the Super Source™ series by participating in the outlining, writing, and field testing of the materials.

Managing Editor: Alan MacDonell
Development Editor: Toni-Ann Guadagnoli
Contributing Writer: Cathy Fillmore Hoyt

Production Manufacturing Director: Janet Yearian
Design Director: Phyllis Aycock
Senior Production Coordinator: Fiona Santoianni
Cover Design: Phyllis Aycock
Text Design: Fiona Santoianni

Line Art: Michel Lee
Composition: Andrea Reider

Copyright © 1998 by
Cuisenaire Company of America, Inc.
10 Bank Street, White Plains, New York 10602

All rights reserved
Printed in the United States of America
ISBN 1-57452-175-6

Permission is granted for limited reproduction of pages from this book for classroom use.
The word CUISENAIRE and the color sequence of the rods, cubes, and squares are trademarks of Cuisenaire Company of America, Inc., registered in the United States and other countries.

1 2 3 4 5 - SG - 02 01 00 99 98

...the Super Source™
Table of Contents

INTRODUCTION 4
OVERVIEW OF THE LESSONS 6
LESSONS
   Investigating Sampling 8
      Snapshot 9
      Color Draw 13
      Even-Steven 17
   Investigating Experimental and Theoretical Probability 22
      True Blue 23
      Cube Cover-up 28
      Give and Take 32
      Collectible Cubes 36
      Dizzy Darts 40
   Investigating Central Tendency 45
      Grab Bag 46
      Geo-Hoops 51
      How High? How Long? 56
      Rocket Launch 60
   Investigating Chance 65
      Freeze before Fifty 66
      Half Chance 70
      Block Path 74
   Investigating Permutations 79
      Mall Madness 80
      Quilt Squares 85
      Starting Five 90
BLACKLINE MASTERS
   Activity Masters 95
   Cube Cover-up Game board 114
   Freeze Before Fifty Spinner 115
   Half Chance Game board 116
   Half Chance Spinner A 117
   Half Chance Spinner B 118
   Quilt Squares Patterns 119
   Quilt Squares Record 120

Using the Super Source™

The Super Source™ Grades 7-8 continues the Grades K-6 series of activities using manipulatives. Driving **the Super Source** is Cuisenaire's conviction that children construct their own understandings through rich, hands-on mathematical experiences. There is a substantial history of manipulative use in the primary grades, but it is only in the past ten years that educators have come to agree that manipulatives play an important part in middle-grade learning as well.

Unlike the K-6 series, in which each book is dedicated to a particular manipulative, the Grades 7-8 series is organized according to five curriculum strands: Number, Geometry, Measurement, Patterns/Functions, and Probability/Statistics. The series includes a separate book for each strand. Each book contains activities in which students use a variety of manipulatives: Pattern Blocks, Geoboards, Cuisenaire® Rods, Snap™ Cubes, Color Tiles, and Tangrams.

Each book contains eighteen lessons grouped into clusters of 3-5 lessons each. Each cluster of lessons is introduced by a page of comments about how and when the activities within each lesson might be integrated into the curriculum.

The lessons in **the Super Source** follow a basic structure consistent with the vision of mathematics teaching described in the *Curriculum and Evaluation Standards for School Mathematics* published by the National Council of Teachers of Mathematics. All of the activities involve Problem Solving, Communication, Reasoning, and Mathematical Connections—the first four NCTM Standards.

HOW THE LESSONS ARE ORGANIZED

At the beginning of each lesson you will find, to the right of the title, a list of the topics that students will be working with. GETTING READY offers an *Overview*, which states what children will be doing, and why, and provides a list of *What You'll Need*. Specific numbers of manipulative pieces are suggested on this list but can be adjusted as the needs of your particular situation dictate. In preparation for an activity, pieces can be counted out and placed in containers or self-sealing plastic bags for easy distribution. Blackline masters that are provided for your convenience at the back of the book are also referenced on this materials list, as are activity masters for each lesson.

The second section, THE ACTIVITY, contains the heart of the lesson: a two-part *On Their Own*, in which rich problems stimulate many different problem-solving approaches and lead to a variety of solutions. These hands-on explorations have the potential for bringing students to new mathematical ideas and deepening skills. They are intended as stand-alone activities for students to explore with a partner or in a small group. *On Their Own* Part 2 extends the mathematical ideas explored in Part 1.

After each part of *On Their Own* is a *Thinking and Sharing* section that includes prompts you can use to promote discussion. These are questions that encourage students to describe what they notice, tell how they found their results, and give the reasoning behind their conclusions. In this way, students learn to verify their own results rather than relying on the teacher to determine whether an answer is "right" or "wrong." When students compare and discuss results, they are often able to refine their thinking and confirm ideas that were only speculations during their work on the *On Their Own* activities.

The last part of THE ACTIVITY is *For Their Portfolio*, an opportunity for the individual student to put together what he or she has learned from the activity and discussion. This might be a piece of writing in which the student communicates results to a third person; it could be a drawing or plan synthesizing what has occurred; or it could be a paragraph in which the student applies the ideas from the activity to another area. In any case, the work students produce *For Their Portfolio* is a reflection of what they've taken from the activity and made their own.

The third and final section of the lesson is TEACHER TALK. Here, in *Where's the Mathematics?*, you as the teacher can gain insight into the mathematics underlying the activity and discover some of the strategies students are apt to use as they work. Solutions are also given when they are necessary and/or helpful. This section will be especially helpful to you in focusing the *Thinking and Sharing* discussion.

USING THE ACTIVITIES

The Super Source is designed to fit into a variety of classroom environments. These can range from a completely manipulative-based classroom to one in which manipulatives are just beginning to play a part. You may choose to have the whole class work on one particular activity, or you may want to set different groups to work on two or three related activities. This latter approach does not require full classroom sets of a particular manipulative.

If students are comfortable working independently, you might want to set up a "menu"—that is, set out a number of activities from which students can choose. If this is the route you take, you may find it easiest to group the lessons as they are organized in the book—in small clusters of related activities that stimulate similar questions and discussion.

However you choose to use **the Super Source** activities, it would be wise to allow time for several groups or the entire class to share their experiences. The dynamics of this type of interaction, where students share not only solutions and strategies but also thoughts and intuitions, is the basis of continued mathematical growth. It allows students who are beginning to form a mathematical structure to clarify it and those who have mastered isolated concepts to begin to see how these concepts might fit together.

At times you may find yourself tempted to introduce an activity by giving examples or modeling how the activity might be accomplished. Avoid this. If you do this, you rob students of the chance to approach the activity in their own individual way. Instead of making assumptions about what students will or won't do, watch and listen. The excitement and challenge of the activity—as well as the chance to work cooperatively—may bring out abilities in students that will surprise you.

USING THE MANIPULATIVES

Each activity in this book was written with a specific manipulative in mind. The six manipulatives used are: Geoboards, Color Tiles, Snap Cubes, Cuisenaire Rods, Pattern Blocks, and Tangrams. All of these are pictured on the title page of this book. If you are missing a specific manipulative, you may still be able to use the activity by substituting a different manipulative. For example, most Snap Cube activities can be performed with other connecting cubes. In fact, if the activity involves using the cubes as counters, you may be able to substitute a whole variety of manipulatives.

The use of manipulatives provides a perfect opportunity for authentic assessment. Watching how students work with the individual pieces can give you a sense of how they approach a mathematical problem. Their thinking can be "seen" in the way they use and arrange the manipulatives. When a class breaks up into small working groups, you can circulate, listen, and raise questions, all the while focusing on how your students are thinking.

Work with manipulatives often elicits unexpected abilities from students whose performance in more symbolic, number-oriented tasks may be weak. On the other hand, some students with good memories for numerical relationships have difficulty with spatial reasoning and can more readily learn from free exploration with hands-on materials. For all students, manipulatives provide concrete ways to tackle mathematical challenges and bring meaning to abstract ideas.

Overview of the Lessons

INVESTIGATING SAMPLING

Snapshot 9
In this activity, students take turns randomly sighting one Snap Cube in a box of 22 Snap Cubes. They use compiled data to predict the contents of the box. Then students make a duplicate box to test their predictions.

Color Draw 13
In this activity, students create random samplings with replacement to predict the color distribution of a bag of Color Tiles. Students analyze recorded data, compare the data to what is known about the possible contents of the bags, and draw a conclusion about the contents of their bag.

Even-Steven 17
Students create random samplings with Snap Cubes to test games for fairness. Students then create their own game and verify its fairness using sampling techniques.

INVESTIGATING EXPERIMENTAL AND THEORETICAL PROBABILITY

True Blue 23
Students play a game in which they pick 2 Color Tiles from a bag containing 3 red and 3 blue tiles. Students predict the outcome of the game, compare experimental and theoretical probabilities to their predictions, and redesign the game to reach a different outcome.

Cube Cover-up 28
Students use Snap Cubes to build a three-dimensional record of the results of multiple dice rolls. They also predict the number of rolls it would take to fill the *Cube Cover-up* game board, and then they compare the theoretical probability to the experimental probability.

Give and Take 32
Students draw Snap Cubes from a bag and record the outcome of their picks. Based on the samples collected, students assess the game's fairness, and then make changes to the contents of the bag in order to produce a fair outcome.

Collectible Cubes 36
Students use Snap Cubes to conduct simulations to find the most likely number of trading card packages that they would need to buy in order to collect complete sets of specialty cards.

Dizzy Darts 40
Students design and color a dart board made from Tangram pieces, find the theoretical and experimental probabilities that a "dart" will land on each color, and discover a fair way to assign point values to the colored sections.

Probability and Statistics, Grades 7-8

INVESTIGATING CENTRAL TENDENCY

Grab Bag 46
Students grab a handful of Cuisenaire Rods and find the mean, median, and mode of the lengths of those rods. Then they modify the contents of the bag and predict the mean, median, and mode of the lengths of the new set of rods.

Geo-Hoops 51
Students toss pipe cleaner hoops onto a Geoboard target, score their tosses for ten trials, and calculate mean, median, and mode. The student who has the higher value in at least two out of the three measures of central tendency wins the game. Pairs combine data to determine the typical score of a *Geo-Hoops* turn.

How High? How Long? 56
Students build Cuisenaire Rods into towers to generate data. Then they record, graph, and analyze the data.

Rocket Launch 60
Students gather data as they sample Tangram triangle rocket launches, and measure the distance they travel. Then students use the data to find the typical distance a Tangram rocket travels.

INVESTIGATING CHANCE

Freeze before Fifty 66
In this game of chance, students use random numbers and operations to be the player who collects the closest number to 50 Snap Cubes without going over.

Half Chance 70
In this activity, two players use spinners to determine the color and number of Cuisenaire Rods to place on a rectangular grid in an effort to cover one-half the grid.

Block Path 74
Students play a game in which they use Pattern Blocks, selected by chance and skill, to build a path that ends closest to the finish line.

INVESTIGATING PERMUTATIONS

Mall Madness 80
Students search to find all possible three-color combinations and permutations using the four different colors of Color Tiles.

Quilt Squares 85
Students work in small groups using Tangram pieces to find all the possible arrangements of colors that can be used to fill a quilt square pattern.

Starting Five 90
Students use Snap Cubes to represent players on a basketball team. Students need to arrange the players in as many different ways as possible to model various situations.

Investigating Sampling

1. Snapshot, page 9   (Snap Cubes)
2. Color Draw, page 13   (Color Tiles or Snap Cubes)
3. Even-Steven, page 17   (Snap Cubes or Color Tiles)

Each of the lessons in this cluster can be used to introduce new material or to reinforce ideas that have already been explored in class. In each lesson, students will need to collect, analyze, and organize data from samplings. Students will also have the opportunity to make predictions, explore fairness and discover the Law of Large Numbers.

1. Snapshot (Exploring hypotheses and predictions)

This activity allows students to make and test hypotheses based on sampling.

In *On Their Own* Part 1 and Part 2, students are expected to make predictions about the number of contents in a box of Snap Cubes based on the results of samplings.

Before the activity, teachers might want to review the basis of making a prediction. This could be done simply by showing the students an empty jar and asking them to predict how many scoops of rice it would take to fill the jar. After recording students' predictions on the board, the teacher should start to fill the jar and stop at the halfway point. The students should then be asked if they would like to change their predictions. After recording their changes, the teacher would finish filling the jar.

This introduction gives students the opportunity to make predictions, and then see that with more information (in this case finding out how many scoops fill half of the jar), their predictions can become more accurate. This might encourage them to take a large sample before making a prediction about the box of Snap Cubes.

2. Color Draw (Exploring sampling with replacement)

This activity gives students a chance to encounter sampling with replacement. Teachers will probably want to present this activity after *Snapshot* so students can discover the similarities and differences between sampling activities that are done with and without replacement.

In *On Their Own* Part 1 and Part 2 students collect and organize data from their samplings. In *For Their Portfolio* students need to decide how sampling could help them to make accurate predictions. In this activity, it is important for students to understand the Law of Large Numbers (discussed in *Where's the Mathematics?*, page 16). This activity suggests using Color Tiles, but Snap Cubes could be used as a substitute.

3. Even-Steven (Exploring fairness)

In this activity, students test different versions of a game in order to explore the concept of fairness. *On Their Own* Part 1 asks students to figure out which of three different sampling-game versions is fair. This activity is extended by *On Their Own* Part 2, in which students are asked to design their own sampling variation so that it too would be a fair game.

Before starting this activity, the teacher may want to introduce the concepts of theoretical and experimental probability. Students may then understand that although a game variation may seem fair, the theoretical probability could prove otherwise. For more information about this see *Where's the Mathematics?* (page 20).

SNAPSHOT

- Sampling
- Analyzing data
- Making predictions

Getting Ready

What You'll Need

Snap Cubes, 20 each of four different colors, per group

Snapshot boxes*, 2 per group, one empty and one containing 22 Snap Cubes in four colors in the following quantities:
  1st color 10 cubes
  2nd color 6 cubes
  3rd color 4 cubes
  4th color 2 cubes

Tape

Activity Master, page 96

Overview

In this activity, students take turns randomly sighting one Snap Cube in a box of 22 Snap Cubes. They use compiled data to predict the contents of the box. Then students make a duplicate box to test their predictions. In this activity, students have the opportunity to:

- collect and analyze data
- make and test hypotheses
- work with ratio and proportion

Other *Super Source* activities that explore these and related concepts are:

Color Draw, page 13

Even-Steven, page 17

* Here's how to make a snapshot box:

1. Cut off a corner of a shoe box so that a single Snap Cube inside the box will be visible but will not fall out through the hole.
2. Fill the box with the Snap Cubes listed above. Remember that each group gets one box with contents and one empty box.
3. Tape the lid securely on the box that contains the Snap Cubes.

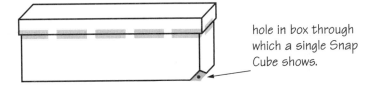

hole in box through which a single Snap Cube shows.

The Activity

On Their Own (Part 1)

> A snapshot freezes a moment in time. A series of snapshots becomes a picture record of an event. Can you use a collection of "snapshots" to find out the contents of a box of Snap Cubes?
>
> - Work in a group. There are 22 Snap Cubes in your Snapshot box.
> - Take turns shaking and tilting the box so that a Snap Cube appears in the opening. Record the color of the cube. Think of each sample as a snapshot of the contents of the box.
> - Keep taking and recording "snapshots" until you think you have enough data to make a prediction.

- Use your data to predict how many of each color there are in the box.
- Be ready to show your data and explain how you arrived at your prediction.

Thinking and Sharing

Have groups share their data and predictions. Create a class chart of each group's predictions.

Use prompts like these to promote class discussion:

- How many times did you take a "snapshot" of the cubes inside the box?
- How did you decide when you had enough data to predict the contents of the box?
- How did you use the data you had collected to figure out how many cubes there were of each color? Did you use any mathematical procedures? Explain.
- All the boxes have the same number and ratio of four colors, but they may have different colors. Is there a way to use the class data to refine your prediction? Explain.
- If you were to repeat the sampling, would you expect the results to be the same? Explain.

On Their Own (Part 2)

What if... you made a second Snapshot box using the number and colors of cubes to match your prediction? If you repeat the experiment with your new box, will the data you collect be close to that collected from the first sampling?

- Work in the same groups, and set up your own Snapshot box. Fill your box with Snap Cubes to match your prediction. Now, tape the lid shut.
- Repeat the experiment from Part 1. Take the same number of "snapshots" that you did with the first box. Record the data.
- Compare the data to that collected from the first box.
- Decide whether you still think the contents of the boxes are the same. Be ready to explain your reasoning.

Thinking and Sharing

Have groups share their predictions and the data collected from both experiments. Record each group's data in a chart.

Use prompts like these to promote class discussion:

- Compare the results of the first and second samplings. How are they the same? How are they different?
- What do the results of the 2nd sampling say about your prediction? Based on the data collected from the second box, do you still believe your prediction is accurate? Why or why not?

- Would it be helpful to compile class data from the 2nd sampling? Explain.

Have groups open the first Snapshot box and compare the contents with their predictions.

- Was your prediction reasonably accurate? What would you do differently next time, if anything, to make a closer prediction?

Suppose that you are writing an article in your school newspaper and you want to find out how students at your school would answer this question: Which would you rather be, very rich or very smart? You want to find out what percentage of students would choose each alternative. Write a summary describing how you would use sampling to gather the data you need. Include the number of samples you would take, how you would make sure the samples were random, and how you would use the data to draw conclusions about students your age.

Teacher Talk

Where's the Mathematics?

In this activity, students explore the mathematical principle called sampling. Statisticians use samples from a large population to make inferences about the population. For example, pollsters interview sample voters to find which candidate the voter population is likely to elect. If the sample size is too small, the results may not be an accurate indication of how the larger population will actually cast their votes. On the other hand, if the sample size is too large, the study becomes costly and time-consuming. In this activity, students investigate appropriate sample sizes for making a prediction.

Some groups may choose to make a prediction after a relatively small sample. Students may theorize that if they take exactly 22 samples, the results should match the contents of the box. If students make a prediction after a small number of samples, encourage them to write down the prediction and take more samples to test the accuracy of the prediction. Some groups may want to make hundreds of observations. Encourage these groups to stop periodically and review the results. At a certain point, they may see that their results are no longer changing and an accurate prediction can be made.

Students who have experience with sampling activities may realize that proportion can be used to arrive at a prediction. If the total number of observations is a multiple of 22 (the target size), calculating proportion will be easier. For example, if the sample size is 88, the sample is 4 times the number of cubes in the box. This means that the number of each color observed is likely to be about 4 times the actual number of that color. Therefore students need only to divide the number for each color by 4 in order to calculate their predictions. For sample sizes that are not multiples of 22, students can solve proportion equations to arrive at a prediction.

$$\frac{\text{(number of red cubes observed)}}{\text{(total sample size)}} \quad \frac{45}{100} = \frac{x}{22} \quad \begin{array}{l}\text{(predicted red cubes)}\\ \text{(total cubes in box)}\end{array}$$

$$45 \cdot 22 = 100x$$
$$990 = 100x$$
$$9.9 = x$$

Since 9.9 is nearly 10, students could conclude that there are 10 red cubes in the box.

Based on prior experience, some students may suggest compiling the data from all groups to aid in making accurate predictions. Others may conclude that this is impossible since the boxes may contain different colors. Students may point out that each box contains the same number breakdown of different-colored cubes. That is, if the cube totals in one box turned out to be 10, 7, 4, and 1, then every box would have 10, 7, 4, and 1. The colors would be different, but the numbers would be the same. From their samples, groups can predict which color was used the most, which was used second most, and so on. Using this information, the groups can compile their data. If students suggest this approach, you may want to organize it on a chart like the one below. The data for each group should be similar to that obtained by your class.

Group	1	2	3	4	5	Totals
1st Color	44	41	23	32	55	**195**
2nd Color	27	22	13	18	28	**108**
3rd Color	20	17	9	11	17	**74**
4th Color	9	8	5	5	10	**37**
Total samples	**100**	**88**	**50**	**66**	**110**	**414**

Students may choose to use proportion and the class totals to check their predictions. For example, using the information from the chart and the method outlined above, students should figure 10 or 11 cubes in the box for the color used most often.

In the second activity, students may be surprised at the limited usefulness of the data from the second sampling. Students may make comments like "You could put the wrong numbers in the box and still get similar results." Some students may suggest compiling the data from both samples. Students should quickly discover that this will not work since the populations sampled (cubes reflecting each group's prediction) may not be the same.

Likewise, students should realize that they cannot compile class data as they did with the first sampling because the groups' predictions are not the same. The second sampling is useful only to the group who conducted the observations.

This activity can help students understand what degree of confidence to place in a sample. Groups who did not make reasonably close predictions may conclude that they needed to base their predictions on larger samples.

Students should be reminded of the sampling and predictions that are made in realistic situations. As students explore these concepts, they may begin to question the reliability of the statistics they read in the newspaper and hear on the news, recognizing that the size of the sample affects the statistics' accuracy and its usefulness.

COLOR DRAW

- Sampling
- Organizing and interpreting data
- Making predictions

Getting Ready

What You'll Need

Paper bags, 1 per group, each containing 12 red, 9 blue, 6 green, and 3 yellow Color Tiles

Paper bags, 1 per group, each containing one of the following combinations of Color Tiles:

    10 Y, 5 G, 3 R, 2 B

    5 Y, 5 G, 5 R, 5 B

    8 B, 6 R, 3 Y, 3 B

    8 B, 8 Y, 2 R, 2 G

Activity Master, page 97

Overview

In this activity, students create random samplings with replacement to predict the color distribution of a bag of Color Tiles. Students analyze recorded data, compare the data to what is known about the possible contents of the bags, and draw a conclusion about the contents of their bag. In this activity, students have the opportunity to:

- experiment using sampling with replacement
- collect and organize data
- use proportional reasoning
- explore the connection between making predictions and the size of the sample

Other *Super Source* activities that explore these and related concepts are:

Snapshot, page 9

Even-Steven, page 17

The Activity

On Their Own (Part 1)

> There are 30 Color Tiles in a bag. The tiles are red, blue, green, and yellow. Can you figure out the number of tiles of each color by checking the color of only one tile at a time?
>
> - Work in a small group. Take turns sampling the tiles in the bag by drawing 1 tile from the bag without looking inside.
> - Each time you draw a tile, record the color and then return it to the bag.
> - Continue sampling the contents of the bag until you are ready to predict the number of each color in the bag.
> - Record your group's predictions, the number of each color picked, and the total number of samples you took.
> - Be ready to explain how you made your predictions.

Thinking and Sharing

Have groups share their predictions about the contents of the bag, the total number of samples drawn, and the number of each color drawn. Explain that each bag has identical tiles, and compile the data from the groups to find the class totals. Allow students to make new predictions based on the compiled data. Then have them look inside their bags to check their predictions.

Use prompts like these to promote class discussion:

- How did you decide how many samples to take?
- How did you use the results of the sampling to make a prediction? Did you use any mathematical procedures? Explain.
- How do your group's results compare with the combined class data? Explain any differences.
- After seeing the combined class data, did you want to change your prediction? Why?
- If you changed your predictions after examining the class data, was your new prediction more accurate than your original prediction?
- Do you think there is any connection between the number of tiles sampled and the accuracy of a prediction? Explain.

On Their Own (Part 2)

What if... your bag contains 20 tiles in one of four possible combinations? Could you use sampling to figure out the contents of your bag?

- Work in small groups. Take turns sampling the contents of your bag. These are the possible combinations of tiles in your bag.

    10 yellow, 5 green, 3 red, and 2 blue

    5 yellow, 5 green, 5 red, and 5 blue

    8 blue, 6 red, 3 yellow, and 3 blue

    8 blue, 8 yellow, 2 red, and 2 green

- Draw 1 tile from the bag without looking inside. Record the color of the tile and then return it to the bag.
- Continue sampling the contents of the bag until you are ready to predict which combination the bag contains.
- Record your prediction, the number of times each color was drawn, and the total number of samples you took.
- Be ready to explain how you used the results of the sampling to reach a decision.

Thinking and Sharing

Have groups summarize their data, share their predictions, and check the contents of their bags.

Use prompts like these to promote class discussion:

- How did you use the results of the sampling to draw a conclusion? Did you use any mathematical procedures to make your decision? Explain.
- How did you decide when you had enough samples to make a decision?
- At what point did you first predict an outcome? Did your prediction change as the experiment continued?

For Their Portfolio

You are on a committee to find out how many students at your school plan to buy tickets to the school play. The information will help you figure out how many performances to schedule. You don't want to ask all of the students in the school if they plan to come to the play. How could you use sampling to make an accurate prediction? Write a summary explaining what you would do to solve the problem.

Teacher Talk

Where is the Mathematics?

Color Draw gives students an opportunity to work with sampling and predicting. In trying to discover the contents of a closed bag, students use the recognized technique of sampling with replacement to learn about the population in which they are interested (in this case, the tiles in the bag). Students are encouraged to draw and replace tiles as many times as they think necessary in order to make a close prediction.

During the first activity, it is not likely that the students will predict exactly what is in the bag. Even after several samples, the data on which their predictions are based may not accurately represent the breakdown of tile colors in the bag. Students should conclude that the larger the sample is, the closer their prediction will be. Some students may point out that some tiles may never be drawn and that some tiles may be chosen over and over again. This possibility becomes quite small when large amounts of data are available, as in the shared class results.

Your may want to organize your class data from the first activity in a chart like this. Your class findings may be similar to these.

		Red	Blue	Green	Yellow	Total samples
Group 1	Sample Results	18	11	12	4	**45**
	Predictions	12	7	8	3	
Group 2	Sample Results	36	28	18	8	**90**
	Predictions	12	10	6	2	
Group 3	Sample Results	38	32	19	11	**100**
	Predictions	11	9	6	4	
Group 4	Sample Results	24	18	13	5	**60**
	Predictions	12	9	7	2	
	Total Results	**116**	**89**	**62**	**28**	**295**

As students make predictions, they naturally apply proportional reasoning to interpret the samples they have drawn. For example, students know there are 30 tiles in the bag, so if 20 out of 60 samples are red, they may conclude that there are 10 red tiles in the bag (20/60 = 10/30).

In the course of the activity, students can observe that proportional reasoning alone does not ensure the accuracy of predictions based on sampling. In fact, for a prediction to be nearly accurate, proportional reasoning must be applied to a sizable sample. Based on the class totals, the use of proportion is likely to yield a very close prediction.

$116/295 \approx 12/30$ $89/295 \approx 9/30$ $62/295 \approx 6/30$ $28/295 \approx 3/30$

Students may realize that the proportions will be easier to calculate if the sample size is some multiple of the total number of tiles in the bag. Based on a total of 30 tiles, students may suggest using a sample size of 60, 90, or 120.

As students compare their group's data with the combined class data, they may find subtle differences. On the sample class chart, Group 1 predicted more green than blue tiles in the bag, because they drew out one more green than blue tile during the sampling. After seeing the combined data, the students in Group 1 may conclude that they need more samples to be sure. Some students may suggest taking at least two or three times as many samples as the number in the bag. Students should see that the larger the sample, the better the chance of making an accurate prediction. This principle is known as the Law of Large Numbers.

In the second activity, students know the contents of the bag are one of four possibilities. Groups will not be able to combine data in this activity since there is no way of knowing which groups have bags with the same contents. Knowing this, students need to take many samples to be sure their predictions are accurate. Since the bag contains 20 tiles, taking samples with a total that is a multiple of 20 will make proportional reasoning easier. Students may suggest that 60 to 100 draws are necessary to be reasonably sure of making an accurate prediction. Most students should agree that 20 or even 40 draws are too few.

As they evaluate the data from the second activity, students may use proportion to arrive at a prediction and then compare their prediction to the possible choices, or they may use a more intuitive method. You may hear statements like "Since we have more yellow than green and not many red or blue, it must be the first combination." Some may eliminate combinations systematically. Students may make statements like: "We have already drawn 20 yellow and only 5 blue, so there is no way the third combination is the right one."

As they work, some members of the group may encourage taking more samples to make sure the proportions among the colors remain the same as the data set grows larger. As students listen to classmates describe how they made their predictions and why they chose a particular combination, students see that evidence-based predicting is different from simply guessing.

EVEN-STEVEN

- Sampling
- Making and testing hypotheses
- Fairness

Getting Ready

What You'll Need

Snap Cubes, 10 each of three different colors per pair

Paper bags, 3 per pair

Activity Master, page 98

Overview

Students create random samplings with Snap Cubes to test games for fairness. Students then create their own game and verify its fairness using sampling techniques. In this activity, students have the opportunity to:

- draw conclusions based on a sample
- collect and analyze data
- make decisions about fairness

Other *Super Source* activities that explore these and related concepts are:

Snapshot, page 9

Color Draw, page 13

The Activity

On Their Own (Part 1)

Steven has designed three versions of a two-player game called Even-Steven. He wants each player to have the same chance of winning so that his game is fair. Which, if any, of Steven's game variations are fair?

- Players take turns drawing two Snap Cubes from a paper bag. Player 1 scores one point if the colors are the same; Player 2 scores one point if the colors are different. The player with the most points after 20 draws is the winner.

- Predict which of the following versions are fair:

   Version 1: The bag holds 1 Snap Cube of one color and 2 Snap Cubes of another color.

   Version 2: The bag holds 2 Snap Cubes of one color and 2 Snap Cubes of another color.

   Version 3: The bag holds 1 Snap Cube of one color and 3 Snap Cubes of another color.

- Decide who is Player 1 and who is Player 2. Now play Version 1. Here's how:

    ◆ Put the Snap Cubes in the bag for Version 1.
    ◆ Take turns drawing two Snap Cubes from the bag.
    ◆ Record the score, and then put the cubes back in the bag.
    ◆ Continue until you have completed 20 trials.

- Now change the contents of the bag. Repeat the activity for Version 2 and Version 3.
- Use your data to decide which variations of the game are fair. Be ready to explain your reasoning.

Thinking and Sharing

Have pairs report their results for Version 1 and indicate whether they believe the version is fair. Create a chart like the one shown below.

Same (Player 1)	Version 1 Different (Player 2)	Conclusion
7	13	Unfair
9	11	Fair
5	15	Unfair

Do the same for Versions 2 and 3.

Use prompts like these to promote class discussion:

- Did the sampling results match your predictions?
- How did you decide whether the version was a fair one? Explain any additional mathematical procedures you may have used.
- Does the compiled class data support the decision your pair made about the fairness of the version? Why or why not?
- Using the compiled class data, which versions, if any, are fair games?
- Are there any other methods, besides sampling, that can be used to evaluate the versions? Explain.

On Their Own (Part 2)

What if... you wanted to design a variation of Even-Steven that uses three colors of Snap Cubes? Use the same rules for scoring as you did in the first three versions. How many cubes of each color would you put in the bag to make a fair game?

- Working with your partner, decide how many Snap Cubes of three colors to put in the bag. Use up to 10 cubes of each color.
- Test the game by making 20 draws and recording your scores.
- If you believe the game is fair, repeat the game at least two more times to collect more data.

- If the game seems unfair, make adjustments to the contents of the bag and try again.
- Be ready to justify why you believe your version is a fair one.

Thinking and Sharing

Have pairs share their versions and display their data.

Use prompts like these to promote class discussion:

- How did you decide how many of each color to put in the bag at first? Did you use any mathematical procedures or strategies to help you decide?
- If you concluded that your first version was not a fair game, how did you decide what changes to make?
- How do you know that your final version is a fair one? Explain.
- What would you expect to happen if a fair version were played ten times?

For Their Portfolio

Steven decides to try another version using 4 red and 2 blue Snap Cubes. Without actually playing the game, describe what you expect the results would be. Do you think the version would be a fair one? Write a summary explaining your reasoning.

Teacher Talk

Where's the Mathematics?

This sampling activity gives students the opportunity to see the relationship between theoretical data and experimental (empirical) data. Students may be surprised that the third version makes *Even-Steven* a fair game. Intuition might indicate that the second version is the fair one since it contains two of each color and most people associate equality with fairness.

Theoretically, in order for a game such as *Even-Steven* to be fair, both players have to have an equally likely chance of winning. That means that the number of possible win outcomes must equal the possible number of no-win outcomes.

Tree diagrams, lists, and matrices provide ways to find out whether one outcome is as likely to happen as another. To analyze the outcomes, cubes of the same color must be distinguished from each other. In this tree diagram for Version 1, B_1 and B_2 represent two blue cubes and R_1 represents one red cube. The tree diagram indicates the six possible outcomes, two that win and four that do not. Version 1 is unfair since the probability of a match is 2/6 or 1/3 and the probability of a non-match is 4/6 or 2/3.

Version 1 Outcomes

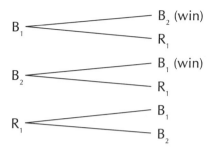

Students may prefer to make an organized list of possible outcomes. Again, numbers are used to distinguish between cubes of the same color. Since each cube in Version 2 can be paired with three other cubes, there are 12 pairs, or permutations. It doesn't matter which cube of the pair is counted first; therefore, half of these are duplicates and can be crossed off the list, leaving six different pairs, or combinations. These combinations are:

Version 2 Outcomes

B_1, B_2 (win)	~~B_2, B_1 (win)~~	~~R_1, B_1~~	~~R_2, B_1~~
B_1, R_1	B_2, R_1	~~R_1, B_2~~	~~R_2, B_2~~
B_1, R_2	B_2, R_2	R_1, R_2 (win)	~~R_2, R_1 (win)~~

The list shows that there are two ways to make a match. The probability of a match is 2/6 or 1/3. Likewise, the probability of a non-match is 4/6 or 2/3. Therefore Version 2 is also unfair because the likelihood of a non-match is more probable than that of a match.

Another way to analyze the possible outcomes is with a matrix—a grid in which each cube in the bag is listed once along the top and once down the side. Here is a matrix showing the outcomes for Version 3. Again, duplicates are crossed off.

Version 3 Outcomes

	B1	R1	R2	R3
B1	X	no win	no win	no win
R1	~~no win~~	X	win	win
R2	~~no win~~	~~win~~	X	win
R3	~~no win~~	~~win~~	~~win~~	X

The Xs in the matrix represent the four impossible outcomes. Same color pairs are marked *win*, indicating that the probability of a match is 3 out of 6, or 1/2. Since the same probability exists for a non-win, Version 3 is the only fair version of the game.

When students try using three colors in Part 2, the only fair version of *Even-Steven* uses the colors in a 9:3:1 ratio. Students who use any of the methods described above may find the correct ratio. Students may use theoretical probability to calculate the probability of drawing a matching pair. First, calculate the theoretical probability of drawing a pair of each of the three colors. To do this, imagine drawing the cubes one at a time. Multiply the chance of drawing a color first by the chance of drawing the same color a second time on the next draw. Then, add to find the probability of drawing a pair of any one of the three colors.

Example: 9 red, 3 blue, and 1 yellow

Chance of drawing red on first draw: 9/13
Chance of drawing another red on second draw: 8/12 or 2/3
(8, because 1 of the 9 reds has been removed. 12, because 1 of the 13 cubes has been removed.)
Multiply to find the combined probability of drawing a red pair: 9/13 x 2/3 = 18/39 or 6/13

Chance of drawing blue on first draw: 3/13
Chance of drawing another blue on second draw: 2/12 or 1/6
Multiply to find the combined probability of drawing a blue pair: 3/13 x 1/6 = 3/78 or 1/26

Chance of drawing yellow on first draw: 1/13
Chance of drawing another yellow on second draw: 0
Multiply to find the combined probability of drawing a yellow pair: 1/13 x 0 = 0

Add: 6/13 + 1/26 + 0 = 36/78 + 3/78 = 39/78 = 1/2

There is a 1 in 2 chance of drawing a pair. Likewise, there is an equally likely chance of not drawing a pair. As students explore making changes to make a fair game, they may be surprised at how "uneven" the colors must be to result in a nearly fair game.

Investigating Experimental and Theoretical Probability

1. True Blue, page 23   (Color Tiles or Snap Cubes)
2. Cube Cover-Up, page 28   (Snap Cubes)
3. Give and Take, page 32   (Snap Cubes or Color Tiles)
4. Collectible Cubes, page 36 (Snap Cubes)
5. Dizzy Darts, page 40   (Tangrams or Pattern Blocks)

The lessons in this cluster give students a variety of opportunities to explore probability. Each activity encourages students to understand the distinct differentiation between experimental and theoretical probability. Students will be required to make predictions, collect and analyze data, and investigate the relationship between experimental and theoretical probability.

1. True Blue (Exploring probability using sampling with replacement)

This is a sampling activity with replacement. The students should have some knowledge about the difference between experimental and theoretical probability before the activity.

In *On Their Own* Part 1 students are expected to figure out the experimental, theoretical and combined probabilities of their samples. Part 2 extends this practice by giving students a desired outcome (1 in 3 chance of winning) and they need to adjust the sample accordingly. *Where's the Mathematics?* (page 26) gives teachers organized lists of the outcomes for Part 1 and Part 2.

2. Cube Cover-Up (Exploring probability using dice rolls)

This lesson gives students a good indicator of the usefulness of theoretical probability when making a prediction. In *On Their Own* Part 1 students use dice rolls to determine where to place Snap Cubes on a game board made up of 36 spaces (one for each dice roll). This activity allows students to see how a change in the number of possible outcomes causes a change in theoretical probability.

In *For Their Portfolio*, students encounter a useful, realistic approach to using probability when they calculate the chances of winning the lottery.

3. Give and Take (Exploring probability using sampling without replacement)

In this activity, students test a game for fairness by sampling tiles from a bag, without replacement. In *On Their Own* Parts 1 and 2, students are to determine what makes a game fair or unfair. Students who have an understanding of ratios will be able to understand how the ratio of colors in a bag affects the probability of the colors drawn from the bag.

In *For Their Portfolio* students are given the opportunity to investigate this activity using sampling with replacement. This activity suggests using Snap Cubes, but Color Tiles could be used as a substitute.

4. Collectible Cubes (Extending probability using sampling with replacement)

This activity provides students with a chance to use probability via sampling with replacement, while modeling a real-life situation. In *On Their Own* Part 1 students are expected to make predictions and organize data. In Part 2 students need to figure out the ratio of cubes in order to create the appropriate simulation. *Where's the Mathematics?* on page 38, offers teachers varied interpretations of sample data.

5. Dizzy Darts (Exploring probability using game results)

In this lesson, students are expected to find both experimental and theoretical probability results based on their Tangram dart board game results.

In *On Their Own* Part 1 students need to calculate the area of a Tangram dart board and they also need to convert probability ratios into percents. In Part 2 students need to assign point values to the colors on their dart boards according to area and probability.

TRUE BLUE

- Experimental probability
- Theoretical probability
- Combined probability

Getting Ready

What You'll Need

Color Tiles, 10 blue and 10 red per pair

Paper bags, 1 per pair

Activity Master, page 99

Overview

Students play a game in which they pick 2 Color Tiles from a bag containing 3 red and 3 blue tiles. Students predict the outcome of the game, compare experimental and theoretical probabilities to their predictions, and redesign the game to reach a different outcome. In this activity, students have the opportunity to:

- make predictions and draw conclusions based on data
- investigate the relationship between theoretical and experimental probability
- make and test hypotheses
- calculate combined probability

Other *Super Source* activities that explore these and related concepts are:

Cube Cover-up, page 28

Give and Take, page 32

Collectible Cubes, page 36

Dizzy Darts, page 40

The Activity

On Their Own (Part 1)

Trina wants to win a goldfish at the carnival. In order for her to win, she needs to pick 2 blue tiles out of the "True Blue prize bag," without looking. If the prize bag contains 3 blue tiles and 3 red tiles, what is the probability of winning the game?

- Working in pairs, put 3 blue and 3 red Color Tiles in a bag. Take turns drawing out 2 Color Tiles. If you draw a blue pair, record a "win." Then replace the tiles.

- Predict the number of wins your team will get if you play the game 40 times. Be ready to explain how you arrived at your prediction.

- Now, conduct 40 trials and record the outcomes. Using your results, write the experimental probability of winning *True Blue*. (The experimental probability is the ratio of wins to total trials).

- Compare the experimental probability to your prediction.
- Find the theoretical probability of winning *True Blue*. Here's how:
    (1) Imagine picking the tiles one at a time. To find the theoretical probability that the first tile will be blue, write the ratio of the number of blue tiles to the total number of tiles in the bag.
    (2) Imagine the first blue tile has been drawn. To find the theoretical probability that the second tile will be blue, write the ratio of the number of blue tiles remaining in the bag to the total number of tiles remaining.
    (3) Multiply the theoretical probabilities to find the combined theoretical probability.
- Compare the experimental probability and the combined theoretical probability of winning *True Blue*. Be ready to explain any differences.

Thinking and Sharing

Have pairs share their experimental probabilities with the class. Combine the data in a chart and find the total of wins and trials for the entire class.

Use prompts like these to promote class discussion:

- How did you make your prediction? Explain any mathematical procedures you used.
- How did your prediction compare to the results of your experiment? Were you surprised by the results? Explain.
- You based your theoretical probability on two separate draws. When finding probability, is there any difference between drawing two tiles at once and drawing them one at a time? Explain your reasoning.
- How did you compare your theoretical and experimental probabilities? What mathematical procedures did you use?
- Find the experimental probability from the class data. Which experimental probability came closer to the theoretical probability: the results from the class as a whole, or the results from your own trials? What would you expect to happen if the class conducted even more trials? Explain.

On Their Own (Part 2)

What if... you wanted to change the game of *True Blue* so that Trina would have a 1 in 3 chance of winning the goldfish? How would you do this?

- Working with your partner, use up to 10 red and 10 blue tiles. Decide how many of each color to put in the bag. Then calculate the theoretical probability of drawing 2 blue tiles.

- Adjust the numbers of red and blue tiles until the theoretical probability of drawing two blue tiles is 1 in 3. Be ready to explain how you decided how many of each color to use.
- Find the experimental probability of winning your game by conducting at least 30 trials.
- Compare the theoretical and experimental probabilities. Be ready to explain any differences.

Thinking and Sharing

Have pairs share how many of each color they used to redesign the game. Discuss the strategies they used to change the chance of a win.

Use prompts like these to promote class discussion:

- What numbers of each color did you start with? Why did you choose these numbers?
- What methods and strategies did you use to make your work easier? Did you organize your work in any special way?
- What mathematical procedures did you use to compare the theoretical probability with the desired outcome of a 1:3 ratio?
- How did the experimental probability compare to the theoretical probability? Explain any differences.

Suppose you play *True Blue* with 3 blue, 3 red, and 3 yellow tiles in the bag. Write a brief summary describing how to calculate the theoretical probability of drawing 2 blue tiles. If you were to conduct an experiment of 40 trials, what results would you expect? Explain your reasoning.

Teacher Talk

Where's the Mathematics?

In this activity, students are given opportunities to find combined probabilities and compare experimental and theoretical probabilities. As they find and compare probabilities, students strengthen their understanding of ratio and fraction concepts.

In the game of *True Blue,* students are likely to expect one result and get another. Students with little experience with probability may predict 20 wins, thinking that there are only two possible outcomes: match and no match. Students with some knowledge of probability may predict 10 wins, thinking that blue-blue is 1 of 4 possible outcomes: blue-blue, blue-red, red-blue, and red-red.

Some students may recognize that outcomes vary based on which blue or red tile is drawn. To make a prediction, they may make an organized list of possible outcomes. In the table below, subscript numbers are used to identify the tiles. For example, the first blue tile becomes B_1. Since it doesn't matter in which order the tiles are drawn, the following table lists all 15 possible combinations. Three of the 15 combinations contain 2 blue tiles, which reduces to a theoretical probability of 1 in 5. Making an organized list of combinations is a good way for students to verify the mathematics involved in calculating combined probability.

(B_1, B_2)	(B_2, B_3)	B_3, R_1	R_1, R_2	R_2, R_3
(B_1, B_3)	B_2, R_1	B_3, R_2	R_1, R_3	
B_1, R_1	B_2, R_2	B_3, R_3		
B_1, R_2	B_2, R_3			
B_1, R_3				

During the activity, students should find a theoretical probability of 3/6, or 1/2, for getting blue on the first draw; and a theoretical probability of 2/5 for getting blue on the second draw. The combined probability (3/6 × 2/5) is 6/30 or 1/5. Based on the theoretical probability, 40 trials will likely yield close to 8 wins, because 8/40 = 1/5.

Drawing the tiles one at a time rather than together will not change the probabilities, as long as the first tile is not replaced before drawing the second one. This fact may not be obvious to students, and they may need to experiment both ways to see if there is an actual difference in results.

Students may use a variety of methods for comparing probabilities. The important point is to determine how close probabilities are. For this purpose, most students will find it easiest to change fractions to decimals before comparing.

The combined class data may look something like the chart below. The total should form a ratio close to the theoretical probability of 1/5. Students may conclude that as the number of trials increase, experimental probability more closely approaches theoretical probability. This idea is often called The Law of Large Numbers.

Sample Class Chart:

Wins

$$\frac{6}{40} \quad \frac{8}{40} \quad \frac{9}{40} \quad \frac{8}{40} \quad \frac{7}{40} \quad \frac{10}{40} \quad \frac{8}{40} \quad \frac{5}{40} \quad \frac{11}{40} \quad \frac{9}{40} \quad \frac{8}{40} \quad \frac{7}{40} = \frac{96}{480} = \frac{1}{5}$$

Trials

As students work to redesign the game, they will have ample opportunities to calculate combined probability. Some may suggest using a computer to calculate probabilities. As they work on their game designs, some students may be satisfied with a ratio close to 1/3; others may search for a combination of Color Tiles that yields an exact theoretical probability of 1/3. Using no more than 10 tiles of each color, the only combinations that result in a theoretical probability of 1/3 are 2 blue:1 red and 6 blue:4 red.

Students who do not have access to a computer or calculator will need to come up with shortcuts, because there are 121 possible combinations of 10 or fewer blue and 10 or fewer red tiles. As they work through possible combinations, students may notice that the number of blue tiles must always be greater than the number of red tiles in order to achieve a theoretical probability of 1 win in 3 trials.

The table below shows the most likely number of red tiles that would have to combine with blue tiles to result in a probability of 1/3. You may want to use this table as a guide when evaluating student work.

Blue Tiles	Red Tiles	Combined Probability of Drawing 2 Blue Tiles	Decimal Equivalence (to the nearest thousandth)
10	7	45/136	0.331
9	6	12/35	0.343
8	5	14/39	0.359
7	5	7/22	0.318
6	4	1/3	0.333
5	3	5/14	0.357
4	3	2/7	0.286
3	2	3/10	0.300
2	1	1/3	0.333

CUBE COVER-UP

- Experimental probability
- Theoretical probability
- Making and testing hypotheses
- Organizing and interpreting data

Getting Ready

What You'll Need

Snap Cubes, about 150 per pair

Dice, one pair of 2 different colors, per pair

Cube Cover-up Game board, page 114

Activity Master, page 100

Overview

Students use Snap Cubes to build a three-dimensional record of the results of multiple dice rolls. They also predict the number of rolls it would take to fill the *Cube Cover-up* game board, and then they compare the theoretical probability to the experimental probability. In this activity, students have the opportunity to:

- investigate the relationship between theoretical and experimental probability
- make and test hypotheses
- collect and interpret data

Other *Super Source* activities that explore these and related concepts are:

True Blue, page 23

Give and Take, page 32

Collectible Cubes, page 36

Dizzy Darts, page 40

The Activity

On Their Own (Part 1)

How many rolls of one pair of dice will you need to place at least one Snap Cube in each of the 36 spaces on the Cube Cover-up game board?

- Using the two colors of your dice, assign one color to represent the top row of dice on the *Cube Cover-up* game board and use the other color to represent the side row of dice on the board.
- Working in pairs, Player A rolls the dice, and Player B records the roll by placing a Snap Cube on the corresponding space on the *Cube Cover-up* game board. If the same dice combination is rolled more than once, Player B should stack cubes on top of one another on that space.
- Predict how many dice rolls you will need to cover each space on the board with at least one cube. Record your prediction before you begin rolling.
- When every space has at least one cube, count the total cubes to find out how many times you rolled the dice. Compare your results to your prediction.

- Create a table to show how many spaces are covered by 1 cube, 2 cubes, 3 cubes, and so on up to 6 or more cubes. And, include a column in your table to show the total cubes needed to cover the board.
- Switch roles and repeat the experiment.
- Be ready to discuss your results, predictions, and tables.

Thinking and Sharing

Have pairs display their tables and explain their results. Create a frequency graph to record total cubes from each trial. Invite pairs to plot the results from their trials on the class graph.

Use prompts like these to promote class discussion:

- How did your actual results compare with your predictions?
- Did your results surprise you? Why or why not?
- Are some combinations easier or harder to roll than others? Explain.
- Examine the chart you and your partner created. Which quantities of stacked cubes (2 cubes on one space, 3 cubes on one space, etc.) occurred most frequently? Why do you think this happened?
- Find the theoretical probability of covering an uncovered space when there are 36 uncovered spaces; 12 uncovered spaces; 4 uncovered spaces; 1 uncovered space. How does the theoretical probability change when more spaces are covered? How does this affect your perception of which spaces are hard to cover?

On Their Own (Part 2)

What if... you picked only 12 spaces on the *Cube Cover-up* game board that you wanted to cover? About how many rolls of the dice would you need to cover those spaces?

- Working with your partner, circle 12 spaces on the *Cube Cover-up* game board.
- Predict how many dice rolls you would need to cover your circled spaces.
- One partner rolls the dice and the other partner records using Snap Cubes.
- As soon as you have covered your chosen spaces, count the Snap Cubes on the board to find the total number of rolls.
- Compare the results with your prediction.
- Repeat the experiment a second time. You may use the same 12 numbers or choose different ones.
- Be ready to discuss your results.

Thinking and Sharing

Have pairs report their predictions and actual number of rolls.

Use prompts like these to promote class discussion:

- How did your actual results compare to your predictions?
- Did your results surprise you? Why or why not?
- Did the particular numbers you chose affect the total number of rolls? Explain.
- How does this experiment compare with a lottery in which people try to choose five or six winning numbers out of a set of numbers?
- Are some numbers "luckier" than others? Explain.
- How did the actual results for this experiment compare to the first experiment of covering all 36 spaces?
- Did the results of the 36-space experiment affect your predictions in this experiment? Explain.

In a lottery, players predict which 5 numbers out of the numbers from 1 to 36 will be drawn out at random. Two friends want to enter the lottery. One friend picks the numbers 6-14-23-29-34. The other friend picks the numbers 1-2-3-4-5. Which friend do you think has a better chance of winning the lottery? Explain your answer.

Teacher Talk

Where's the Mathematics?

In this activity, students are given opportunities to investigate the relationship between theoretical and experimental probability using dice and Snap Cubes. As they roll dice and record their results, they strengthen their understanding of how change in the number of possible outcomes causes change in theoretical probability.

Students will probably need anywhere from 120 to 180 cubes to fill their boards. The following table may be similar to those made by your students.

Spaces with:	Trial 1	Trial 2
1 cube	3	4
2 cubes	7	8
3 cubes	12	8
4 cubes	4	7
5 cubes	4	3
6 or more cubes	6	6
Total cubes on board	132	130

Predictions and actual results may vary. Some students may predict that it will take only 36 rolls to cover the board. Others may recognize from the start that the final spaces may take many rolls to cover. Most will underestimate the number of rolls it will actually take to cover the board.

As students work they may assume that the sums with a greater number of cubes are easier to get than those that are not filled. They may make statements like "Two 4s is the easiest combination to roll because there were 5 cubes on that space" and "You have to be really lucky to roll two 3s; it was the last space I covered."

Students may choose to test such statements by finding the theoretical probability. Then they may make statements like "The chance of rolling any certain combination is still 1 in 36," or "There's no such thing as luck; the dice rolls are random."

Students may observe that the theoretical probability of rolling an uncovered combination decreases after each new space is covered. The theoretical probability of hitting 1 of 36 uncovered spaces is 36/36 or 1; that is, the probability of covering a space is certain. For 12 uncovered spaces, the theoretical probability is 12/36 or 1/3; for 4 uncovered spaces, 1/9; and for 1 uncovered space, 1/36.

When you begin to create a frequency graph of students' results on the board, have students suggest a range or use a range from 95 to 180 as shown below. Your class frequency graph may look something like this:

			X		
		X	X		
		X	X	X	
	X	X	X	X	X
X	X	X	X	X	X
95-109	110-123	124-138	139-153	154-168	169-180

Based on these results, students may conclude that they would have a good chance of covering all 36 spaces after 150 dice rolls. Some students may point out that it would be possible to cover all the spaces with 36 dice rolls. Others may emphasize that it would also be possible to make thousands of dice rolls and not cover all the spaces. Encourage students to use correct probability vocabulary such as likely, unlikely, probable, possible, and so on.

During the activity or the discussion, some students may insist that they are "lucky" and that they can roll certain combinations easily. By repeating the activity several times, students may gain greater confidence in using theoretical probability, rather than luck, as a guide to making predictions. Through repetition, students may discover that with a greater number of trials the results will be closer to a theoretical model.

In the second activity, students choose 12 dice combinations to roll. Since most students will want to cover their 12 numbers in as few turns as possible, pairs may again discuss which combinations are easier to roll. Students may be surprised to find that the number of dice rolls needed to cover 12 particular spaces is not much less than the number needed to cover all 36 spaces. Based on their observations during the first activity, students who see the similarity between the two activities may make closer predictions on the second activity.

The second activity is similar to a lottery in which players try to predict which numbers will be drawn from a set of numbers. However, the activity differs from a lottery because the emphasis is on the number of rolls and not on predicting which 12 numbers will be rolled first.

GIVE AND TAKE

- Experimental probability
- Sampling
- Fairness

Getting Ready

What You'll Need

Snap Cubes, 15 of one color, 15 of a second color, and 20 of a third color, per group

Paper bags, 1 per group, each containing the above quantities of Snap Cubes

Activity Master, page 101

Note to teacher: Prepare bags in advance by putting the specified set of Snap Cubes in each. On each bag write the three colors of Snap Cubes that are inside. Do not write the quantity of each color.

Overview

Students draw Snap Cubes from a bag and record the outcome of their picks. Based on the samples collected, students assess the game's fairness, and then make changes to the contents of the bag in order to produce a fair outcome. In this activity, students have the opportunity to:

- explore probability concepts
- draw conclusions based on a sample
- explore concepts of fairness
- analyze data to test the design of a game

Other *Super Source* activities that explore these and related concepts are:

True Blue, page 23

Cube Cover-up, page 28

Collectible Cubes, page 36

Dizzy Darts, page 40

The Activity

On Their Own (Part 1)

> Give and Take is a game for 3 players. The object is to be the first player to collect 10 Snap Cubes of a chosen color. Without looking in the bag, can you tell whether each player has an equal chance of winning?
>
> - Work in groups of three. Each player should choose the color he or she wants to collect from the choices listed on the bag.
> - Without looking, take turns drawing a cube from the bag. Give the cube to the player who is collecting that color. The first player to collect 10 cubes is the winner.
> - When the game ends, record the color that won and the number of cubes collected for each color.
> - Play several times, switching colors each time. Continue to record data.
> - Without looking into the bag, decide whether the game is fair. Be ready to explain your reasoning.

Thinking and Sharing

Without revealing the contents of the bags, invite groups to discuss whether they believe the game is fair. Collect data from each group. You may wish to display the class data in a chart like the one below.

How many cubes per color?										
1st Color	5	6	8	7	5	10	5	8	7	10
2nd Color	6	10	5	4	6	6	7	5	10	7
3rd Color	10	7	10	10	10	9	10	10	8	9

Use prompts like these to promote class discussion:

- What makes a game fair? Can you think of any unfair games that people like to play? Name them. Why do you think they are unfair?
- How did you decide whether this game was fair? Did you perform any mathematical procedures to reach a decision? Explain.
- How many games do you think you'd have to play to be sure the game was fair or unfair?
- How does seeing the class data affect your conclusion about the fairness of the game?

On Their Own (Part 2)

> **What if...** you wanted to change the game of Give and Take to make it fair? You still do not know the exact contents of the bag. What would you change about the game? How can you be sure your version of the game is fair?
>
> - Work in the same groups of three. Using the data collected in the first activity, decide how many cubes would change the contents of the bag enough to make Give and Take a fair game.
> - You may add or subtract cubes, but you may not look in the bag. Keep a record of the changes you make.
> - Now play the game as you did before. Record which color wins and the number of cubes collected for each color.
> - Play the new game several times. Continue making changes, playing the game, and collecting data until you are sure the game is fair.
> - Be ready to explain what changes you made and how you know your game is fair.

Thinking and Sharing

Have groups share the changes they made in the game and the data they collected. Then ask students to check the contents of their bags. Have them count the total number of cubes and the number of each color.

Use prompts like these to promote class discussion:

- Based on the contents of the bag, is your game fair? How did you decide?
- How did you use your game results to decide how to change the game?
- Is there a relationship between the data you collected and the number of cubes of each color in the bag? Explain.
- Does the likelihood of picking a particular color change as more cubes of that color are picked? Why or why not?
- Does the total number of cubes in the bag affect the results? Explain.

Suppose you put the cubes back into the bag after each turn. How would the game change? Write a brief summary describing the changes in your results. Be specific about how you would decide if the game was fair.

Teacher Talk

Where's the Mathematics?

In this activity, students explore experimental probability as they test a game for fairness. As students draw out cubes, they sample the cubes in the bag. After playing the game several times, students are able to use the collected data to help predict the number of cubes of each color in the bag. As they use data to evaluate their games, students strengthen their understanding of how sampling is used to make decisions about a larger population.

Give and Take is a game that may be changed to favor a certain color to win or to give all players equally likely chances of winning. The game goes quickly, allowing students to collect a lot of data in a short amount of time. The game starts with 15 cubes of one color (red), 15 of a second color (white), and 20 of a third color (blue). Blue has an advantage since there are one-third more blue cubes than there are of either of the other colors. On average, ten blue cubes will be collected in the same amount of time it takes to collect seven or eight red or white cubes.

Once students have seen that blue seems to win repeatedly, they will probably determine that the game is unfair. Students are often concerned with the issue of fairness. Most games are fun because they are fair, meaning that each player has an equal chance of winning or losing. People do, however, play games that are unfair. Casinos and carnivals have many "unfair" games in which the odds of winning are stacked against the player. Students may point out that people play unfair games because they hope they will be lucky and beat the odds.

Making a class chart allows the class to pool its data and see trends that might not appear in a small data set. From the data, students should easily see that there must be more blue cubes in the bag than either red or white. However, analyzing the data to determine the ratio of red to white to blue is more difficult. At first, students may think that the ratio based on the cubes in their data should be a close match to the ratio based on the cubes in the bag. This would be the case if the cubes were replaced each time (as in *For Their Portfolio*). Sampling without replacement, however, has an effect on the likelihood of a particular color being picked. The most "popular" color will probably have more cubes in the bag than the students predicted.

In Part 2, watching groups make changes to make the game fair provides a good opportunity to assess whether students understand how the ratio of colors affects the probability of the color drawn and the outcome of the game. Students may be surprised to find that even a minor change can greatly effect the play of the game. As students evaluate their games, they may determine that the game is fair when each color wins an equal number of times. Others may focus on the ratios of the colors drawn during each game. These students may be satisfied with producing a close game, regardless of which color wins.

After students count the cubes in the bag, they should be able to see a relationship between data collected and the number of cubes of each color in the bag. Students may make statements like "There was a good chance that blue would win as long as there was more blue in the bag" and "Once we took out some of the blue, then red and white won more often." Some students may conclude that the trick to making the game fair is to use the same number of each color.

Students should recognize that the total number of cubes in the bag does not matter as long as the numbers are equal and there are at least 10 of each color. If too many cubes are taken from the bag, it could become impossible for any color to win. If too many cubes are added, students may have difficulty shaking or stirring the cubes in the bag to distribute the colors randomly. Through experimentation, students may decide that it is best to add as well as subtract cubes from the bag to find a balance of color.

Students who work on the *For Their Portfolio* activity may suggest drawing a cube and replacing it as a better method for sampling the contents of the bag. This method would result in data that is closer to the actual contents of the bag. With replacement, a given color is always equally likely to be picked, no matter how often it has already been picked. On the other hand, in Parts 1 and 2, a color that has been picked several times becomes *less* likely to be picked, because there are fewer cubes of that color in the bag.

COLLECTIBLE CUBES

- Simulation
- Sampling
- Analyzing data

Getting Ready

What You'll Need

Snap Cubes, approximately 20 per pair, at least 6 of one color and 2 each of 6 different colors

Paper bags, 1 per pair

Activity Master, page 102

Overview

Students use Snap Cubes to conduct simulations to find the most likely number of trading card packages that they would need to buy in order to collect complete sets of specialty cards. In this activity, students have the opportunity to:

- use objects to model a real-world situation
- organize and analyze data
- make predictions and draw conclusions based on data

Other *Super Source* activities that explore these and related concepts are:

True Blue, page 23

Cube Cover-up, page 28

Give and Take, page 32

Dizzy Darts, page 40

The Activity

On Their Own (Part 1)

> The Jock 'n' Rock trading card company puts a special hologram card in each package of trading cards. There are 6 different hologram cards. The company sends an equal number of each kind of hologram card to every store. How many packages of trading cards do you think you would need to buy to have a good chance of getting all 6 hologram cards?
>
> - Working with a partner, predict how many packages you think you would need to buy in order to get all 6 hologram cards.
> - Perform a simulation. Here's how:
>   - Put 6 different-colored Snap Cubes in a paper bag. Without looking, pick a cube from the bag and record the color. Return the cube to the bag and shake the bag to mix up the Snap Cubes.
>   - Continue picking cubes, recording the colors and returning them to the bag until you have picked all 6 colors. Record the number of picks you made.
> - Run the simulation at least 3 times.

- Decide how many packages of trading cards you would need to buy to have a good chance of getting all 6 hologram cards.
- Be ready to explain why you chose that number of packages.

Thinking and Sharing

Have pairs share their predictions, the results from their trials, and the number of packages they believe is the best number to buy based on their trials. Create a frequency distribution graph to combine the data from the trials.

Use prompts like these to promote class discussion:

- How did you make your prediction? Did you use any mathematical procedures to calculate the best number to buy? Explain.
- How did you organize the data you collected during each trial?
- How did you use the data from your combined trials to choose the best number of packages to buy?
- After looking at the class' graph, would you choose a different number of packages to buy? Explain your decision.
- What do you think would happen to our class graph if we conducted and recorded more trials?
- When you conduct the simulation, why is it important to replace each cube before drawing out another cube?

On Their Own (Part 2)

What if... the Jock 'n' Rock trading card company decides to put one or the other of two special edition cards in each of its trading card packages? Suppose the company prints 3 times more of the "Jock of the Month" card than the "Rocker of the Month" card. How many packages of trading cards do you think you would need to buy to have a good chance of getting both special edition cards?

- With your partner, make a prediction based on the changes in the situation.
- Design a simulation using Snap Cubes and a paper bag to model the events in this situation.
- Run your simulation at least 3 times.
- Use data from your trials to decide how many packages of cards you would buy.
- Be ready to explain your predictions and simulations, and why you chose that number of packages.

Thinking and Sharing

Have pairs share their predictions, the results from their trials, and the number of packages they believe is the best number to buy based on their trials. Again, create a frequency distribution graph to combine the data from the trials.

Use prompts like these to promote class discussion:

- How did you make your prediction? Did you use any mathematical procedures to calculate the best number to buy? Explain.

- What was your simulation? Why did you choose to test your prediction in this way?

- How did you use the data from your combined trials to choose the best number of packages to buy?

- After looking at the class' graph, would you choose a different number of packages to buy? Explain your decision.

- Compare the class' graphs from both simulations. How are they the same? How are they different?

- What do you think will happen to the graph for this simulation if we complete more trials? Explain your reasoning.

For Their Portfolio

The trading card company wants to increase its number of hologram cards from 6 to 25. A package of trading cards sells for $1.75. Using this information, write a letter to the company president encouraging or discouraging the company from making this decision and give reasons to support your request.

Teacher Talk

Where is the Mathematics?

The process of using actual objects to model a situation is a powerful tool that students can use to analyze real-life situations. This activity gives students a chance to learn the value of a simulation. The data from the simulation reflect the actual outcomes of the realistic situation.

Since the Snap Cubes are randomly picked from the bag and then returned to the bag, each cube has an equally likely chance of being picked each time. The fact that a cube has already been picked once or twice does not make it less likely to be picked the next time. Given this, the range for the number of picks it takes to choose all six cubes can vary considerably. Students may observe that the smallest number of picks possible is 6. However the chances are low that any pair will get this result. It is also possible, but unlikely, that a pair of students will not pick all six cubes during the class period.

Students may use a variety of methods to organize the data from each trial. Some may list the color names and make tally marks. Others may tally the total cubes drawn and list a color name only when it is drawn for the first time. Students may benefit from a discussion of which recording method was easiest to use.

In theory, purchasing 14 to 15 packages of trading cards is likely to give a complete set of six hologram cards. Students may use mathematical means to arrive at an answer. As they consider the mathematics in the situation, they may notice that the probability of drawing a new color constantly changes. For example, the probability of getting a new color on the first draw is 6/6, or 1, but on the next draw, the probability is 5/6. The final color has a 1/6 chance of being drawn. Students may make statements like "Since there is a 3 out of 6 chance to get the fourth color, which reduces to 1 out of 2, let's say it will take 2 turns to get the fourth color." Using this reasoning, they may create the following equation: 6/6 + 6/5 + 6/4 + 6/3 + 6/2 + 6/1 = 14.7 picks.

The following frequency distribution graph may be similar to that made from your students' data:

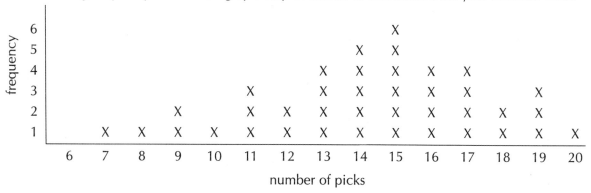

Students may interpret the information from the graph and their trials in a number of ways. Some may answer the question of how many packages of trading cards will be needed by giving a range of numbers. Other students may describe an "average" number of picks, some actually calculating the mean of the values. Others may find the median (middle value) or mode (most used value). Any of these could be considered reasonable responses.

In the second simulation, students must use a 3:1 ratio to make a simulation that fairly reflects the situation. One possible simulation would be to put 3 of one color and 1 of another in the bag. Again, a drawn cube must be replaced before another cube is drawn.

In theory, purchasing 5 to 6 packages of cards should give a good chance of getting the pair of special edition trading cards. Students may make statements like "You should be able to get the Jock of the Month card in 1 or 2 tries" and "Since the probability of getting the Rocker of the Month card is 1 in 4, it will probably take about 4 tries to get it." Students may notice that if they draw the Rocker of the Month card first, the likelihood of getting both cards in two tries is very high. Some may create an equation to make this prediction: 4/1 + 4/3 = 5 1/3.

The class's frequency distribution graph for the second simulation may look something like this:

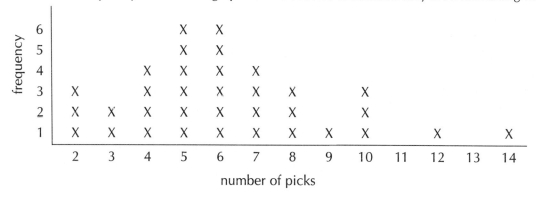

As students compare the graphs, they may notice the bell-like shapes formed by the data. They may suggest that as more data are collected, the bell shapes will become more uniform. Based on this idea, they may suggest filling in the probable outcomes of further trials.

DIZZY DARTS

- Experimental probability
- Theoretical probability
- Ratio
- Percentage
- Area

Getting Ready

What You'll Need

Tangrams, 2 sets per pair

Crayons or markers

Small objects such as a dime, a bean, or a small ball of clay to be used as darts, 2 per pair

Calculators (optional)

Activity Master, page 103

Overview

Students design and color a dart board made from Tangram pieces, find the theoretical and experimental probabilities that a "dart" will land on each color, and discover a fair way to assign point values to the colored sections. In this activity, students have the opportunity to:

- create and compare a variety of polygons
- calculate the ratio of one area to another
- write part-to-whole ratios using fractions and percents
- compare experimental and theoretical probabilities

Other *Super Source* activities that explore these and related concepts are:

True Blue, page 23

Cube Cover-up, page 28

Give and Take, page 32

Collectible Cubes, page 36

The Activity

On Their Own (Part 1)

If you drop "darts" onto a Tangram dart board, how can you find the probability of landing on each of the colors on the board?

- Working in pairs, make a polygon using 1 set of Tangram pieces. Trace the polygon onto paper. Include the outline of each Tangram piece. This is your Tangram dart board.

- Color the Tangram dart board using as few colors as possible. Make sure that the sections that share sides are different colors.

- Find the *theoretical* probability of a dart's landing on each color. Here's how:

  ◆ Using the small Tangram triangle as a unit of measure, find the area covered by each color and the area of the whole dart board. Then write a ratio like this for each color:

  $$\frac{\text{area of color section(s)}}{\text{area of dart board}}$$

- Now find the *experimental probability* of a dart's landing on each color. Here's how:
  - Drop a dart onto the Tangram dart board and record the color it lands on.
  - Perform 20 trials. Do not count trials that miss the dart board. If your dart lands on the line between two sections, record the color at the centermost point of your dart.
  - Write a ratio like this for each color to show the experimental probability of hitting each section.

    $$\frac{\text{number of times the dart hits the color}}{20 \text{ (total number of trials)}}$$

- Now convert each probability ratio into a percent.
- Make a table to summarize the results of your work. Be ready to explain any differences that you found between the theoretical and experimental results.

Thinking and Sharing

Have pairs show their dart boards and discuss the theoretical and experimental probability of landing on each color.

Use prompts like these to promote class discussion:

- How did you decide what shape dart board to build and how to color it?
- Did your probability experiment turn out as you thought it would? Explain.
- Compare the theoretical and experimental probabilities for hitting each color. Are there differences? If so, why?
- If you repeat the probability experiment using the same dart board, do you think the results will be the same? Explain.
- What would you expect to happen to the results of the probability experiment if you increase the number of trials to 100? Explain.
- If you rearrange the pieces on your dart board without changing the color of any piece, will the probabilities change? Explain.
- How are the percents helpful in this activity?

On Their Own (Part 2)

What if... **you wanted to assign points to the different colors on your dart board? How would you do this?**

- Use two complete sets of Tangram pieces to design a new dart board. Trace your Tangram shape onto paper, and remember to include an outline of each Tangram piece.

- Color your Tangram dart board using 4 colors. Make sure that pieces that share sides are different colors.
- Compare the areas of the colors. Then assign each color a point value from 1 to 16 points in a way that you think is fair.
- Now find the theoretical probability of landing on each color. Use these probabilities to decide whether you have assigned the points fairly. Make adjustments if necessary.
- Make a chart showing each color, its point value, and the theoretical probability of landing on the color with a dart.
- Be ready to explain the relationship between the points you have assigned to each color and the theoretical probability of landing on that color.

Thinking and Sharing

Have pairs show their dart boards and charts. Discuss the theoretical probability of hitting each color.

Use prompts like these to promote class discussion:

- How did you go about designing your dart board?
- How did you decide what Tangram pieces to include in each colored section?
- How did you decide how many points to assign to each color on the dart board?
- How did you use theoretical probability to assign points fairly? Explain.
- What relationship did you find between the point values and the theoretical probability?

Suppose the theoretical probability of hitting the color red is 1/2. During an actual game, a player hits red 7 out of 20 times. Write a paragraph comparing the theoretical and experimental probabilities. If you were to conduct another experiment of 20 trials, what results would you expect? Explain your reasoning.

Teacher Talk

Where's the Mathematics?

In this activity, students are given opportunities to explore spatial and numerical relationships within the context of a probability investigation. As they find probabilities, students strengthen their understanding of the concepts of area, ratio, fraction, and percent.

Students can create many different dart boards using their Tangram pieces. Here are some examples:

Students may discover that all Tangram dart boards can be colored with either two or three colors. As they experiment, they may realize that by moving pieces, they can reduce the number of colors.

Some students may talk about probability informally using statements like "There's so much blue, there is no way you can land on red." Students may express an understanding of the relationship of probability to ratio, fraction, and percent through statements like "Since half the shape is red, you have a 50% chance of hitting red." Expressing probability using percents may help students make comparison statements like "You have a four times better chance of hitting red than green because 50% is four times 12½%."

In this activity, students base theoretical probability on the area of each color. Any polygon made from one set of Tangram pieces will have an area equal to 16 small Tangram triangles.

In the polygon below, it would take three small triangles to cover the red pieces, five to cover the blue, and eight to cover the green.

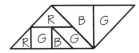

This table, based on the polygon above, may be similar to those created by your students.

Theoretical Probability			Experimental Probability		
Color	Ratio	Percent	Color	Ratio	Percent
Red	3/16	18 3/4%	Red	4/20	20%
Blue	5/16	31 1/4%	Blue	5/20	25%
Green	8/16	50%	Green	11/20	55%

Students may observe that the experimental probabilities may change if the experiment is repeated, but the theoretical probabilities remain the same as long as the area of each color does not change. Even rearranging the pieces will not change the theoretical probabilities. Students may conclude that theoretical probabilities are usually close to the experimental probabilities and can be used as a guide for predicting what will probably happen. Some students may realize that if an experiment contains many trials, the experimental probability is more likely to equal the theoretical probability. Students may make statements like "Theoretical probability tells what will probably happen; experimental probability tells what did happen."

As students design their dart boards in Part 2, they may observe that small extensions on concave polygons may be difficult to hit with a "dart." Students may question whether the experimental probabilities for hitting sections on concave and convex polygons will differ. They should be encouraged to investigate this idea by building a convex target, coloring it, and then moving one or two small pieces to create a concave target. The coloring scheme should remain the same. Students can then find the experimental probability of hitting each color using both targets and compare the results. The theoretical probability for the two targets remains the same as long as the area of each color and the total area of the target remain the same.

Students may use a variety of strategies to assign points. Although there is no single "correct" method, the general idea will be to assign *more* points to areas that are *less* likely to be hit. Some students may apply inverse proportion. To justify their work, students may make statements like "Since green is one half the area of red, it should be worth twice as many points." Some may use percent to assign points: "If green takes up 25% of the space, it should be worth 75% of the points."

This sample dart board with points assigned and the table below may be similar to those made by your students.

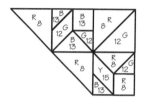

Color	Theoretical Probability	Points Assigned
Red	16/32	8 points
Blue	6/32	13 points
Green	8/32	12 points
Yellow	2/32	15 points

Investigating Central Tendency

1. Grab Bag, page 46 — (Cuisenaire Rods)
2. Geo-Hoops, page 51 — (Geoboards)
3. How High? How Long?, page 56 — (Cuisenaire Rods or Pattern Blocks)
4. Rocket Launch, page 60 — (Tangrams)

The activities in this cluster give students practice in the use of measures of central tendency and in the analysis of data. The lessons are ideal for reinforcing the concepts of mean, median and mode.

1. Grab Bag (Exploring central tendency)

This lesson could be used as an introduction to mean, median, and mode. In *On Their Own* Part 1, students try to find the typical length of a Cuisenaire Rod in a handful of rods taken from a bag. Part 1 contains detailed instructions telling students how to find each measure of central tendency for their handfuls of rods. In *On Their Own* Part 2, the contents of the bag of rods changes and students are asked to make predictions based on the changes.

Throughout the activity, students search to find the meaning of "typical." This topic is discussed in greater detail in *Where's the Mathematics?* on page 50.

2. Geo-Hoops (Extending central tendency)

In this activity, students calculate the mean, median and mode of their *Geo-Hoops* game scores. It is assumed that students have had previous experience with measures of central tendency. In *On Their Own* Part 2, pairs combine their data from Part 1 and find the typical number of points scored on a turn.

Where's the Mathematics? (page 54) discusses the meaning of "average" or "typical" when talking in terms of central tendency.

3. How High? How Long? (Extending data analysis)

In this activity, students test their physical skills while using Cuisenaire Rods. The activity provides good practice in graphing and analyzing data.

In *On Their Own* Part 1 students count the number of Cuisenaire Rods they can build into a tower before the tower falls. In Part 1 and Part 2 students will need to know how to create a frequency graph to record their results.

For Their Portfolio challenges students to create their own experiment, record their data, and then compare their results to the other trials. This activity suggests using Cuisenaire Rods, but Pattern Blocks could be used as a substitute.

4. Rocket Launch (Exploring data analysis)

In this activity, students gather data from sample Tangram triangle rocket launches in order to find the typical distance a rocket travels. In *On Their Own* Parts 1 and 2, students are asked to create a frequency graph. Teachers may want to review or introduce methods of graphing so students can find the best way to organize and consequently analyze their data.

Where's the Mathematics? (page 62) also discusses some of the terms the teacher may want to introduce or reinforce, such as "outlier," "range," and "normal distribution curve."

GRAB BAG

- Measures of central tendency
- Analyzing data

Getting Ready

What You'll Need

Cuisenaire Rods, 1 set per pair

Paper bags, 1 per pair

Activity Master, page 104

Overview

Students grab a handful of Cuisenaire Rods and find the mean, median, and mode of the lengths of those rods. Then they modify the contents of the bag and predict the mean, median, and mode of the lengths of the new set of rods. In this activity, students have the opportunity to:

- investigate mean, median, and mode
- draw conclusions about a data set
- make and test hypotheses

Other *Super Source* activities that explore these and related concepts are:

Geo-Hoops, page 51

How High? How Long?, page 56

Rocket Launch, page 60

The Activity

On Their Own (Part 1)

> Hector insisted that anytime he picked rods out of the class storage box, there would always be a blue rod in his pick. He therefore thought that blue must be the average length for a rod. What do you think? Suppose you reach into a bag and grab a handful of Cuisenaire Rods. What is the typical length of a rod in your handful?
>
> - Each color of Cuisenaire Rod is a different length. The white rod is the shortest at 1 centimeter long; the orange rod is the longest at 10 centimeters long. Note: You may find it helpful to make a chart of rod colors and their corresponding lengths.
>
> - Work with a partner. Put one set of Cuisenaire Rods in a paper bag.
>
> - Player 1 should grab a handful of rods and place them on the desktop, recording the number of each color of rod in the handful.
>
> - Find and record the median length of the handful. Here's how:
>
>   ◆ Organize the rods from longest to shortest. Find the middle rod. If you have an even number of rods, there will be two rods in the middle. If the middle rods are different colors, the median is halfway between their lengths. If the two rods are the same color, the length of that color is the median.

- Find and record the mode length of the handful. Here's how:
  - Sort the handful into piles by color. The pile with the most rods is the color that represents the mode. You can have more than one mode.
- Find and record the mean length of the handful. Here's how:
  - Find the sum of the lengths in the handful. Divide the sum by the number of rods in the handful. Round to the nearest tenth.
- Put the rods back into the bag and have Player 2 grab a handful. Find the mean, median, and mode of the lengths in the handful.
- Now combine the data from both handfuls of rods. Add the rods you grabbed to those in your partner's handful and find the combined mean, median, and mode.
- Be ready to explain what the measures tell you about the typical lengths of rods in a handful.

Thinking and Sharing

Invite pairs to share their work. Create a chart to display each pair's results.

Use prompts like these to promote class discussion:

- Which measure (median, mode, or mean) best represents the typical length of a rod from a handful of Cuisenaire Rods? Explain your reasoning.
- What does the mode tell you about the handful? Why is the mode sometimes very different from the mean and median?
- What does the median tell you about the handful of rods?
- How is the mean useful in describing the typical length of a rod?
- Can you use the mean from both trials to find the combined mean? Explain.
- Which set of data (Player 1, Player 2, or the combined trials) is most useful in describing the typical length of a rod? Why?
- Do you notice any patterns occurring in the class data as a whole?

On Their Own (Part 2)

What if... the contents of the bag were changed? Suppose two orange, two blue, and two brown Cuisenaire Rods were removed from the bag. How would this affect the typical length of a rod in a handful of Cuisenaire Rods?

- First predict the results of your experiment. Predict the median, mode, and mean of a handful of Cuisenaire Rods. Be ready to explain your predictions.
- Take turns grabbing a handful of Cuisenaire Rods from the bag.
- For each handful, find the median, mode, and mean of the lengths of the rods.
- Now combine the data from both handfuls and find the median, mode, and mean of the lengths of the combined rods.
- Be ready to explain the results of your experiment and prediction.

Thinking and Sharing

Have pairs share their predictions and the results of their combined trials. Compile the predictions on a class chart. Use prompts like these to promote class discussion:

- How did you go about making your predictions?
- How did the data you collected compare to your predictions?
- Was it easier to predict the median, mode, or mean? Why?
- What do you notice about the class's predictions? How are they the same? How do they differ?
- Were you surprised by any of the results of your experiment? Explain.
- Compare your data to that gathered during the first activity. Which measure—the mean, median, or mode—was affected most by the removal of the rods? Why?

The sports writer of a newspaper wrote that a baseball pitcher typically throws 92 miles per hour. What does this statistic really mean? What information would help you to understand the statistic? Write a letter to the sports writer explaining what kind of information readers need to know in order to understand this statistic.

Teacher Talk

Where's the Mathematics?

Students investigate the meaning of the three measures of central tendency: mean, median, and mode. As they analyze the lengths in a handful of Cuisenaire Rods, they explore what kinds of information are needed to make statistics meaningful.

Each pair of students begins by putting a full set of Cuisenaire Rods in a paper bag. The set should include 4 orange, 4 blue, 4 brown, 4 black, 4 dark green, 4 yellow, 6 purple, 10 light green, 12 red, and 22 white rods. The results of the activity will not be affected greatly if two or three smaller rods are missing from the set.

The following set of rods may be similar to a handful taken by your students.

Rods in Handful	Total length by color
2 orange	20 cm
4 blue	36 cm
3 brown	24 cm
2 dark green	12 cm
2 yellow	10 cm
1 purple	4 cm
1 light green	3 cm
15 total rods	109 total cm

Mode, Median and Mean of Handful

Mode = 9 cm

Median = 8 cm

Mean = 7.3 cm

The following data sets may be similar to those collected by your students in the first activity.

	Student A	Student B	Combined	Student A	Student B	Combined
Median (cm)	6	6.5	6	7	7	7
Mode (cm)	8, 1	9	8	9	9	9
Mean (cm)	5.6	6.2	5.8	6.9	6.7	6.8
Number of Rods	19	12	31	19	18	37
Total Length (cm)	107	74	181	132	121	253

Students may associate the word *average* with the mean. Actually, all three measures of central tendency can be called an *average* or a *measure of the center* of a set of data. Each measure of central tendency helps you understand how the data in a set are distributed.

As students investigate the median, you may hear statements like, "The median is the middle length so about half of the rods in the handful are bigger and half are smaller" and "The median tells you the middle but it doesn't tell the biggest and smallest pieces in the handful." It is possible that all the rods in the handful are the same length. Without additional knowledge, the median gives a narrow perspective of the data.

Students may be surprised to find that the mode often varies widely from the mean and the median. They may also have difficulty understanding in what situations the mode becomes a useful piece of information. Two real-life examples are T-shirt sizes or tire sizes; students may think of other examples. When the same mode occurs in several handfuls, students may use the information to ask, "Why do we tend to grab more of the longer rods than the shorter ones?" and "Why is the mode usually a longer length of rod even though there are many more short rods in the bag than long ones?"

Students are accustomed to finding the mean and usually refer to it as the average. This custom is followed widely by newspapers and even textbooks, although both the median and the mode are also measures of average. Students learn early on how to add up all the elements and divide by the number of elements, but they often have trouble understanding how the mean represents the set of data. Students may observe that as the median is the middle rod, the mean is the middle of the numbers themselves.

Arguments can be made for any of the measures of central tendency to best represent the average length in a handful of Cuisenaire Rods. As students explore the ideas, they may reach the conclusion that it is most helpful to have more than one of the measures in order to understand the data. They may also conclude that it is helpful to know the range and the number of rods in the handful.

Students may be surprised to find that you can't average averages. To help students understand why it doesn't work, have them discuss the following situation: A baseball player gets only 5 at bats in the first half of the year but gets 4 hits for an .800 batting average. During the second half of the year the player gets 80 hits out of 250 at bats for a .320 average. Averaging the averages gives the player a .560 batting average for the year, but does this average really represent his hitting ability? His actual batting average for the year would be .329 (84 hits ÷ 255 at bats = .329).

The combined trials should yield the most useful information in describing the typical rod length in a handful of Cuisenaire Rods simply because the analysis is based on a larger pool of data. Some students may be concerned that the measures are not accurate because the size of students' hands varies, and so does the number of rods they can grab. Actually, this is one of the reasons why the mean, median, and mode are useful measures: They allow statisticians to compare samples that are not equal in size. Students may suggest compiling data from other groups. If students try this, they may find that there is a point at which new data actually adds nothing new to their understanding of the situation. Likewise, statisticians limit the number and size of the samples they gather.

In the second activity, students will probably predict a lower mean, median, and mode than they found in the first activity. With fewer long rods in the bag, students may find that the mode is very difficult to predict and it may be much lower than the median and mean.

The following data sets may be similar to those collected by your students in the second activity.

	Student A	Student B	Combined	Student A	Student B	Combined
Median (cm)	4	3	3.5	6	6.5	6.3
Mode (cm)	4, 2	1	4, 2, 1	3	7	7, 3
Mean (cm)	4.0	2.7	3.4	5.7	6.3	7.2
Number of Rods	12	11	23	14	14	28
Total Length (cm)	48	30	78	80	121	201

GEO-HOOPS

- Measures of central tendency
- Graphing and analyzing data

Getting Ready

What You'll Need

Geoboard, 1 per pair

Pipe cleaners, five 12" lengths, per pair

Activity Master, page 105

Overview

Students toss pipe cleaner hoops onto a Geoboard target, score their tosses for ten trials, and calculate mean, median, and mode. The student who has the higher value in at least two out of the three measures of central tendency wins the game. Pairs combine data to determine the typical score of a *Geo-Hoops* turn. In this activity, students have the opportunity to:

- compare measures of central tendency
- collect and analyze data
- explore the meaning of "average"

Other *Super Source* activities that explore these and related concepts are:

Grab Bag, page 46

How High? How Long?, page 56

Rocket Launch, page 60

The Activity

On Their Own (Part 1)

Emily invented a game called Geo-Hoops. Geo-Hoops is played much like "Horseshoes," except that instead of ringing a horseshoe around a stake, players toss hoops onto Geoboard pegs. After several rounds, players calculate the mean, median and mode, and the winner is the player with the higher score in at least two of the three statistical measures. How do you think the scoring method will affect the outcome of the game?

- Working with your partner, make five hoops out of pipe cleaners. Here's how:
  (1) Overlap the ends.   (2) Twist them.   (3) Wrap the ends around the hoop.
  Each hoop should be about 3 inches in diameter.

- Place a Geoboard on a flat surface. Mark a throwing line 8 inches away from the board. When it is your turn to toss the hoops, your throwing hand must not cross the line.

- Decide who will go first. Each player gets to toss 5 hoops on a turn. Toss the hoops one at a time, and leave the hoops on the board after each toss. You score points for every Geoboard peg you ring. The pegs are scored as follows:

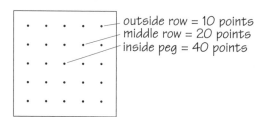
outside row = 10 points
middle row = 20 points
inside peg = 40 points

- Hoops that lean on a peg but do not surround it do not count. If more than one hoop surrounds a peg, count full score for the bottom hoop and half score for the upper hoop(s).

Scoring Examples:

Score all four pegs:
10+10+10+20=50

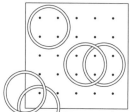

Right hoop counts full for all four pegs:
10+10+20+20=60

Left hoop counts full for two left pegs (40+20=60) and half for right overlapped pegs (10+10=20)

Total score = 50+15+60+60+20=205 points.

Bottom hoop counts 10.
Overlapping hoop counts 5.

- Each player gets 10 turns. Record the scores for each turn.
- When both players have finished 10 turns, calculate the mean, median, and mode of the scores. (If the middle two scores are different numbers, the median is the number halfway between those scores. If there is more than one mode for a set of data, count the higher one.)
- The player who has the highest 2 out of 3 measures (mean, median, and mode) wins.
- If a tie makes it impossible to declare a winner, both players should take another turn. Calculate the new mean, median, and mode. Continue until a player wins.
- Play two games of Geo-Hoops.
- Be ready to discuss your results.

Thinking and Sharing

Have students share their results. Make a chart showing the measures of central tendency for a few pairs' ten turns. Circle the winning measures for each set of results.

Use prompts like these to promote class discussion:

- What was the highest score you earned on a turn? the lowest?

- What throwing strategies did you use to ring the most points possible?
- What do you notice about the data from these pairs? Do you see any patterns?
- Even though the game winner was not determined by the total number of points, was there a relationship between winning and scoring the highest point total? Explain.
- Is it possible to win the game and not have the highest mean? Why or why not? Support your answer with examples.

On Their Own (Part 2)

> **What if...** you wanted to find the typical number of points scored on a turn in a game of Geo-Hoops? How could you use the data you have collected to find out?
>
> - Combine the data from both players for each of the games you played.
> - Calculate the mean, median, and mode of the combined data.
> - Be ready to explain how many points are scored on a typical turn.

Thinking and Sharing

Have pairs share their answers and explain their reasoning. Make a list of the mean, median, and mode of the combined data for each pair.

Use prompts like these to promote class discussion:

- For this game, which statistical measure—mean, median, or mode—do you think best represents the points scored in a typical turn?
- Why was it important to combine the data from both players to find the typical score?
- What can you learn from the class list of typical scores? Would it be helpful to find the mean, median, and mode of the list? Why or why not?

The Math Wiz game company wants to sell the Geo-Hoops game in stores. The marketing department needs feedback from students who have already played the game. Write a letter to the company explaining what you think students can learn from playing Geo-Hoops.

Teacher Talk

Where's the Mathematics?

In this activity, students examine the mean, median, and mode of the scores in a set of *Geo-Hoops* turns. As students calculate and compare measures of central tendency they strengthen their understanding of how these measures can be used to describe a result as "typical" or "average."

The scores for a single turn may range from 0 to a possible, but highly improbable, 350. Students may use a variety of throwing strategies. If an argument about the legality of a throw cannot be handled through the rules as stated, taking the turn over is the best course and an acceptable way to resolve conflict for most students. Allow students to make additional rules about how to define a legal throw if necessary.

Students will find that most of the time the player with the higher point total wins. The mean is directly related to the point total since the players complete the same number of turns. The player with the higher point total, then, will always have the higher mean. However, it is possible to win and not have the higher mean, as shown in the game below in which Player 1 wins.

	Scores from 10 turns	Total	Mean	Median	Mode
Player 1	60 30 60 30 40 40 0 10 60 10	340	34	(35)	(60)
Player 2	30 50 55 30 0 20 10 70 30 55	350	(35)	30	30

In the second activity, students are asked to find the typical score of a *Geo-Hoops* turn. The combined data from the game above would yield the following results:

Combined scores from 20 turns	Total	Mean	Median	Mode
60 30 60 30 40 40 0 10 60 10 30 50 55 30 0 20 10 70 30 55	690	34.5	30	30

Students should recognize the need to combine data to form a larger sample. By combining data from more than one player, students pool different skill-levels. Some students may decide that even 20 turns is not a large enough sample. Students may want to combine data from more than one game or from several groups to obtain more accurate results.

Arguments can be made for any of the measures of center to best represent the typical score. As students explore the ideas, they may reach the conclusion that it is most helpful to have more than one of the measures in order to understand the data. They may also conclude that it is helpful to know the range of the scores.

Although students are accustomed to using the mean to represent the average in a set, some students may raise the question, "How can the mean represent the scores when the mean isn't one of the scores?" As students explore the question, they may conclude that the mean represents the center of the total of the scores just as the median represents the middle score on the list.

This activity also gives student practice in calculating median using an even set of numbers. Students should realize that the median falls between the fifth and sixth numbers in a set of 10 numbers. After putting the values in order, they will need to find the mean of the fifth and sixth numbers, or determine the number that would fall halfway between them. If the fifth and sixth numbers are the same, the median has the same value as those numbers.

Students may observe that while the mean and median are usually relatively close in number, the mode is sometimes much higher or lower. For this reason, some students may decide that the mode is not the best measure of a typical score. Other students may argue reasonably that if most players seem to get one score more often than any other, that score actually is the typical score.

Some students may realize that in this activity you can find the combined mean of two sets of scores by averaging their means. This works only when the data sets contain the same number of scores (in this case 10 trials each). However, students should be reminded that if they needed to find the mean of the scores of two games that did not contain the same number of trials, they would need to add the individual scores and divide by the total number of scores.

Generally, students should be cautioned against "averaging" averages. Students can gain valuable information from the class list by comparing and contrasting the measures listed. However, to perform calculations, the class needs to return to the original game scores.

HOW HIGH? HOW LONG?

- Normal distribution
- Graphing and analyzing data

Getting Ready

What You'll Need

Cuisenaire Rods, 1 set per pair
Activity Master, page 106

Overview

Students build Cuisenaire Rods into towers to generate data. Then they record, graph, and analyze the data. In this activity, students have the opportunity to:

- investigate concepts of central tendency
- collect and analyze data
- create a frequency graph

Other *Super Source* activities that explore these and related concepts are:

Grab Bag, page 46

Geo-Hoops, page 51

Rocket Launch, page 60

The Activity

On Their Own (Part 1)

> **How many white Cuisenaire Rods do you think you can build onto a tower before the tower falls?**
>
> - Work with a partner. Find a level surface to use as a base for your tower.
> - Take turns stacking white Cuisenaire Rods. Place only one rod at a time. Do not hold the tower while putting a new rod on top. You can straighten the tower between turns. Keep building until the tower falls.
> - Record the number of Cuisenaire Rods in the tower. Do not count the final rod in the stack unless it remains in place for at least three seconds.
> - Repeat the experiment 20 times.
> - Create a frequency graph to display your data.
> - Be ready to explain how your data can be used to find out how many white Cuisenaire Rods you and your partner can typically stack.

Thinking and Sharing

Have pairs display their frequency graphs and discuss their results. Compile the data to create a class frequency graph.

Use prompts like these to promote class discussion:

- What do you notice about the shape of the data on your graph?
- What was the greatest number of rods you were able to stack? How does this number relate to the other data on your graph?
- What do you think would happen to your pair's graph if you were to complete more trials? Why?
- Suppose you and your partner decide to build one more tower. What number of rods do you think you could stack? Explain your reasoning.
- How does your graph compare to the class graph? How are they the same? How are they different?
- What is the *typical* or *usual* number of rods which students in this class can successfully stack? How did you use the class data to choose a number?

On Their Own (Part 2)

> **What if...** you wanted to find out how far, with one breath, you can blow a white Cuisenaire Rod along a smooth surface?
>
> - Work with a partner. Find a smooth, flat surface.
> - Mark a starting place for the white Cuisenaire Rod. Decide how you will measure the distance it will travel.
> - Blow the white Cuisenaire Rod along the smooth surface. Experiment a few times to figure out the best angle to blow from.
> - Now, work with your partner to complete 20 trials, 10 trials each.
> - Record your data on a frequency graph as before.
> - Using your data, what distance do you think a "typical" student could move a white Cuisenaire Rod with one breath? Be ready to explain your thinking.

Thinking and Sharing

Have pairs display their frequency graphs and discuss their results. Compile the data to create a class frequency graph.

Use prompts like these to promote class discussion:

- What is meant by a "typical" student?
- Based on your graphs, how far can a typical student blow a white Cuisenaire Rod with one breath? Explain how you used the data to choose this distance.

- How does the shape of this graph compare to the shape of the graph you made for your white-rod towers?
- How does your graph compare to the class graph? After seeing the class graph, would you change the distance a "typical" student can blow a white Cuisenaire Rod?
- What do you think would happen to the class graph if more trials were conducted?
- Would finding the mean, median, and mode of the data be helpful? Why or why not?

Create your own experiment. Decide which ability you would like to test using Cuisenaire Rods. Work with your partner to complete however many trials you think would be sufficient. Record your data and compare it to the other trials. Write a brief summary of your results.

Teacher Talk

Where's the Mathematics?

In this activity, students perform experiments with Cuisenaire Rods, collect and compile data, and develop strategies for describing the data. In both activities, students have the opportunity to recognize a normal distribution curve and see how it can be used to define what is "typical."

As they work on both activities, students may want to exclude trials that did not turn out the way they had planned. For instance, if one partner accidentally shakes the table when stacking rods, the pair may want to start over. You may want to discuss with the class that the purpose of the activity is to find out what typically or usually happens. Since shaking the table is a common occurrence, the trial gives the pair important data for their graph. Emphasize that if students were to exclude anything but the best trials, their data would not be representative.

As they look at the graphs of their classmates and the compiled class graph, some students will recognize the familiar shape of the normal distribution curve. The normal curve will be more pronounced on the compiled class graph than on the pairs' graphs. It is important to remind pairs that their results from this activity were based on only 20 trials. Collecting larger samples of data would provide more reliable information. Students may conclude that if they were to complete more trials, their graphs would more closely resemble the class graph.

The following frequency graph may be similar to those generated by two students working together.

```
                        X
            X   X   X   X
        X   X   X   X   X           X
X       X   X   X   X   X   X   X                           X
---------------------------------------------------------------
7   8   9   10  11  12  13  14  15  16  17  18  19
```

Students are used to focusing on the greatest achievable height or distance. But in these activities, a record-setting achievement is just one of the twenty numbers on the graph. Instead, students are asked to find the point on the graph where most data are clustered or centered. Students may notice that the record-setting marks, both high and low, often stand alone. You may want to introduce the term *outlier* to describe an individual piece of data that has an unusual value. There are no rules about what makes a piece of data an outlier. An outlier is always defined by the statistician's judgment.

Students may notice important differences in their graphs. Although each graph probably has a cluster of data creating the curve shape, the *range* of the data may differ. The range is the interval from the lowest value to the highest value in a set of data. A narrow range produces a steeper curve, as in the sample graph below:

```
                              X
                              X
                   X          X
         X         X          X
         X         X          X
   X     X         X          X      X
   X     X         X          X      X
_____
8   9    10   11   12   13   14   15
```

Students may draw many interesting conclusions about the meaning of "typical" as it is used in these two activities. Some may equate "typical" with "average." These students may decide that the best way to choose a typical number is to average (find the mean of) the collected values. You may want to point out that the mean is only one statistical measure to determine what is average and that there are many meanings of average. Some students may decide that the typical value must be the number at the highest point of the curve. Others may look for a number that is at the center of the data. Some students may suggest "finding the number that's in the middle or that all the other numbers are centered around." As students develop their understanding of the meaning of the word *typical*, they may realize that there are many legitimate ways to summarize data and that no one way is best for every set of data.

Students with prior statistical experience may suggest finding the mean, median, and mode of the data set. These statistical measures provide landmarks for a data set, but by themselves they do not create an accurate picture of a set of data. For example, the mean of the data set above is 11.15, the median is 11, and the mode is 12. Although students can easily calculate these values, they may have difficulty understanding what these measures reveal about the data set and which measure identifies the typical value. Some students may note that without the knowledge of the range and whether a data set has any outliers, the mean, median, and mode can give a false impression about what is typical.

ROCKET LAUNCH

• Measures of central tendency
• Analyzing data

Getting Ready

What You'll Need

Tangrams, 1 set per pair

Inch rulers, 1 per pair

Activity Master, page 107

Overview

Students gather data as they sample Tangram triangle rocket launches, and measure the distance they travel. Then students use the data to find the typical distance a Tangram rocket travels. In this activity, students have the opportunity to:

- investigate concepts of central tendency
- collect and analyze data
- create a graph
- conduct experiments to test predictions

Other *Super Source* activities that explore these and related concepts are:

Grab Bag, page 46

Geo-Hoops, page 51

How High? How Long?, page 56

The Activity

On Their Own (Part 1)

Keia just got a new pool table. She noticed that when one ball hits another ball most of the momentum of the first ball is transferred to the second ball, causing it to move. She wants to see if the same idea will work with Tangrams. Make a Tangram rocket and launcher to simulate Keia's pool table. Using momentum, how far can you make the rocket travel?

- Work with a partner. One person will launch the rocket; the other person will measure the distance it travels.

- Find a smooth surface. Make a small mark on the surface and position the edge of a Tangram square next to the mark. Place the rocket (small Tangram triangle) and the launcher (medium Tangram triangle) as shown below.

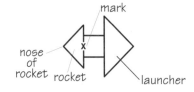

- To launch the rocket, hold the square firmly in place; then slide the launcher several inches away from the square and back again so that you strike the square sharply. The rocket will move away from the square as the momentum transfers.
- Measure the distance from the small mark you made on the surface to the nose of your rocket. Record the measurement. Round your measurement down to the nearest whole inch.
- Repeat the experiment until each player has launched the rocket 20 times.
- Create a frequency graph to display your combined data from 40 launches.
- Be ready to explain the typical distance a small triangle Tangram rocket travels.

Thinking and Sharing

Have pairs display their frequency graphs and discuss their results. Compile the data to create a class frequency graph.

Use prompts like these to promote class discussion:

- How would you define the word *typical*?
- What do you notice about the shape of the data on the graphs?
- What are the lengths of your longest and shortest launches? How do these numbers relate to the typical distance you have determined that a rocket travels?
- What do you think would happen to your graph if you were to conduct more trials? Why?
- Suppose you were going to launch one more rocket. What distance would you predict the rocket would travel? Why?
- How does this activity relate to Keia's pool table? What are the similarities/differences?

On Their Own (Part 2)

What if... the rocket were a medium Tangram triangle and the launcher were a large triangle? What do you predict would be the typical *distance* your new rocket would travel?

- Working with your partner, compare the sizes of the rocket and the launcher in the first activity. Then compare the sizes of the new rocket and its launcher.
- Each person should conduct one sample launch to "feel" how the change in pieces may affect the results of a launch.
- Discuss your findings and make a prediction. Be ready to explain how you arrived at your prediction.
- Conduct 40 launches, 20 by each partner. Measure the distance traveled as before, rounding down to the nearest whole inch.
- Create a frequency graph of your data. Using your data, find the typical distance the rocket traveled.
- Compare your results to your prediction. Be ready to explain any differences.

Thinking and Sharing

Have pairs share their predictions, display their graphs, and explain their results. Compile the data to create a class graph.

Use prompts like these to promote class discussion:

- Did you use any mathematical formulas to help make a prediction? Explain.
- Based on your graph, what was the typical distance the medium triangle rocket traveled? Explain how you chose this number.
- How did the typical distance compare to your prediction?
- Compare your graph to the class graph. How are the graphs the same? How are they different?
- What methods, other than a frequency graph, could you use to analyze your data? Explain.
- Suppose you wanted to increase the distance the medium triangle rocket travels. What could you do?

Two students perform 40 launches using a small Tangram triangle for the rocket and a large Tangram triangle for the launcher. The greatest distance traveled was 44 inches; the smallest distance was 6 inches. To find the typical distance their rocket traveled, they averaged 44 and 6 by adding them and then dividing by 2. Do you think this method is a good one for finding the typical value in a set of data? Write a brief summary explaining your reasoning. Use examples from your experiments to support your answer.

Teacher Talk

Where's the Mathematics?

In this activity, students perform an experiment with Tangram pieces, collect and compile data, and develop strategies for describing the data. Students have the opportunity to recognize how the distribution of data can be used to define what is "typical."

Because students are trying to launch the rocket as far as possible, they may want to exclude trials in which something goes wrong. Students should be reminded that a typical launch may include something going wrong and therefore all trials should be recorded. After the students have completed a few trials, you may want to have a class discussion about whether there are kinds of problems that would justify a "do-over." Emphasize that if students exclude all but their best trials, their data will not represent the typical distance that a Tangram rocket will travel.

As they look at the graphs of their classmates and the compiled class graph, some students will recognize the familiar shape of the normal distribution curve. Other students may describe the data as clumped together at a point or points. Students may be surprised to find some graphs showing more than one curve, indicating more than one typical result. This may occur if one partner's typical launch distance is consistently higher (or lower) than the other partner's (see Sample B on the next page).

The normal curve will be more pronounced on the compiled class graph than on the pairs' graphs. Pairs may conclude that if they were to complete more trials, their graphs would more closely resemble the class graph or demonstrate the normal curve.

The following frequency graphs may be similar to those generated by the pairs in your class.

Sample A

```
            x
            x
        x x     x
        x x x x x
      x x x x x x   x
x x x x x x x   x x
x x x x x x x x x x x x     x           x
─────────────────────────────────────────────
1 2 3 4 5 6 7 8 9 10 11 12 13 14 15 16 17 18 19 20
```

For Sample A, students may report typical launch distance as 3 to 6 inches.

Sample B

```
        x
        x x
        x x         x x
      x x x x     x x x
x x x x x x x x x x x
x x x x x x x x x x x x x         x
─────────────────────────────────────────────
1 2 3 4 5 6 7 8 9 10 11 12 13 14 15 16 17 18 19 20
```

For Sample B, students may report two typical distances, 4 to 5 inches and 9 to 10 inches.

Sample C

```
                        x
                      x x
                    x x x
                  x x x x
                x x x x x
        x x x x x x x x x
x     x x x x x x x x x         x         x
─────────────────────────────────────────────
1 2 3 4 5 6 7 8 9 10 11 12 13 14 15 16 17 18 19 20
```

These students were able to develop the skill necessary to launch the rocket 11 to 12 inches consistently. For Sample C, students may conclude that "the other half of the curve is missing."

As students compare their record-setting marks, both high and low, to the center of the data, they may conclude that certain unusual values are not typical of the set. You may want to introduce that the formal name for one of these values is an *outlier*—an individual piece of data that has an unusual value. There are no rules about what makes a piece of data an outlier. An outlier is always defined by the statistician's judgment.

Students may notice important differences in their graphs. Although each graph may have one or more clusters of data, the *range* of the data may differ. The range is the interval from the lowest value to the highest value in a set of data.

Students may define *typical* as the usual outcome, the average result, or what happens most often. Some students may be surprised that there is not a particular mathematical procedure to determine what is typical.

As students choose a typical value from their data set, some may decide that the typical distance must be the number at the highest point of the curve. Others may use mathematical procedures, both formal and informal, to calculate the typical value. Some pairs may find the mean of the set of data. Others may find the median, concluding that the middle value is best representative of the center of the data. Still others may assume that the distance that occurs most often, the mode, is the typical value.

As students develop their understanding of the meaning of the word *typical,* they may realize that there are many legitimate ways to summarize data and that no one way is best for every set of data. Students may discover that while the mean, median, and mode are helpful tools, one measure alone can give a false impression about a data set. For example, for Sample C, the mean to the nearest tenth is 9.4, the median is 10, and the mode is 12. Standing alone, none of these values adequately describes the shape of the data in the sample.

In the second activity, students compare the relative sizes of the Tangram pieces to make a prediction about the distance a larger rocket will travel. In both activities, the launchers are twice the size of the rockets; however, the rocket will tend to travel somewhat farther in the second activity since it is easier for students to apply greater force to the larger piece. Students may find that the larger pieces result in a greater range with a greater number of outliers. Students should recognize that to increase the distance the rocket travels, the launcher must transfer greater momentum. This could be accomplished by increasing the launcher's size or increasing the force applied to the launcher.

Investigating Chance

1. *Freeze Before Fifty, page 66* (Snap Cubes)
2. *Half Chance, page 70* (Cuisenaire Rods)
3. *Block Path, page 74* (Pattern Blocks)

The activities in this cluster give students practice in probability using game formats. Each lesson has elements of chance and skill and each gives students the opportunity to develop number sense, spatial visualization, and game strategy.

1. Freeze Before Fifty (*Exploring chance*)

In this activity, students spin a spinner and roll a die to determine the number of Snap Cubes they may add to their collection. The object of the game is to "freeze" before collecting more than fifty Snap Cubes.

To play the game, each pair of students needs a number cube marked 2, 2, 2, 3, 3, and 4. If a blank die is not available, teachers can use stickers to change some of the numbers on a regular die, or they can use number cards which students can draw from paper bags.

In *On Their Own* Part 1 the randomness of the spin and the die roll give students exposure to the elements of chance. The ability to freeze when they are close to fifty gives students some control over the game.

Where's the Mathematics? (page 69) gives a chart for teachers who may want to discuss combined probability with their students. This section also provides lists of the possible outcomes from Parts 1 and 2.

2. Half Chance (*Exploring game strategies*)

This is a game in which the choice between two spinners determines which and how many Cuisenaire Rods students should place on a game board. Theoretical probability plays an important part in this activity. Students should be encouraged to try to understand the strategy behind choosing between the two spinners.

In *On Their Own* Part 1, students are given the opportunity to strengthen their strategic thinking skills as they develop game strategies based on probability and spatial reasoning. Their understanding is assessed in *For Their Portfolio* as they write a letter giving advice on the best strategy to use.

In *Where's the Mathematics?* (page 73), teachers will find the theoretical probabilities for each variation of spinner choices.

3. Block Path (*Extending chance*)

In this game, players use Pattern Blocks in an attempt to build a path that ends closest to the finish line on a paper track. The activity contains an element of chance because a dice roll determines how many blocks students add to their paths. However, students need to use strategic skills and spatial visualization when choosing which Pattern Block shape to place.

If students play several games of *On Their Own* Part 1, they will have a chance to develop estimation skills based on spatial reasoning. When considering rule changes in Part 2, students will need to have an understanding of probability in order figure out how to reorganize their game strategies.

Where's the Mathematics? (page 77), is rich with information regarding the theoretical and combined probabilities that students can determine from *On Their Own* Part 2.

FREEZE BEFORE FIFTY

- Chance
- Number sense
- Game strategies

Getting Ready

What You'll Need

Snap Cubes, 100 per pair

Freeze Before Fifty Spinner, page 115

Number cube, marked 2, 2, 2, 3, 3, and 4, 1 per pair

Standard die, 1 per pair

Activity Master, page 108

Overview

In this game of chance, students use random numbers and operations to be the player who collects the closest number to 50 Snap Cubes without going over. In this activity, students have the opportunity to:

- develop strategies based on probability and number sense
- use estimation and mental math
- develop strategic thinking skills

Other *Super Source* activities that explore these and related concepts are:

Half Chance, page 70

Block Path, page 74

The Activity

On Their Own (Part 1)

> **Freeze Before Fifty is a game for 2 players. The object is to collect as close to 50 Snap Cubes as you can without going over. Can you find the winning strategies?**
>
> - Work with a partner. Place 100 Snap Cubes where both players can reach them.
> - Both players roll the 2-2-2-3-3-4 number cube once and collect that number of Snap Cubes. These are your "starting numbers."
> - Decide who will go first. On your turn, spin the *Freeze Before Fifty* spinner and roll the number cube to get an operation and a number. If you spin addition, add the number you roll to the number in your collection. (On your first turn, this number would be your starting number). If you spin multiplication, multiply the number you roll by the number of cubes you already have. Collect that number of cubes and add them to your pile.
> - Take turns spinning and rolling to increase your collections.
> - If you reach more than 50 Snap Cubes on any turn, return all your cubes to the pile and roll a new starting number. On your next turn, start building a new collection of cubes.

- When you think you are as close to 50 as you can get without going over, say "freeze" when your turn comes. Do not add more cubes. The other player gets up to two turns to get closer than you are to 50. The other player can stop at any time.
- Once one player freezes and the other completes his or her final turns, the game is over. The player with the number of cubes closer to 50 without going over wins. If there is a tie, there is no winner, and players should start the game over.
- Play several games of Freeze Before Fifty; take turns going first. Be ready to discuss your strategies for winning.

Thinking and Sharing

Invite pairs to talk about their games and describe some of the strategies they used.

Use prompts like these to promote class discussion:

- Is Freeze Before Fifty a game of chance, of skill, or of both? Explain.
- How did you decide who would go first? Do you think the player who goes first has an advantage? Explain.
- Is it more likely that you would add 4 cubes to your collection or multiply your collection by 2? How do you know?
- How did you decide when to freeze?
- What strategies did you try as you played the game? Did your strategies work?

On Their Own (Part 2)

What if... you change the range of numbers that you could roll? How would this change your game strategies?

- Start the game in the same way. Roll the number cube to choose a starting number.
- Then, roll both dice, the number cube (marked 2 to 4) and the standard die (marked 1 to 6), and find the sum of your rolls.
- Now spin the spinner and add or multiply your total roll by the number of cubes in your collection. Collect that number of Snap Cubes and add them to your pile.
- Continue playing (as you did in Part 1) until there is a winner.
- Play several games of this version of Freeze Before Fifty.
- Be ready to explain how changing the range of possible numbers affected your game strategies.

Thinking and Sharing

Invite pairs to discuss this version of *Freeze Before Fifty*. Ask students about the similarities and differences of the two games.

Use prompts like these to promote class discussion:

- Which version of the game did you like better? Why?
- Which version required more skill?
- How did the range of numbers that could be rolled in this version compare to the range from the first version?
- How did changing the range affect your game strategies?

For Their Portfolio

Suppose you are playing *Freeze Before Fifty* with a die marked with the numbers 1, 1, 1, 2, 3, and 4. How do you think this change would affect the play of the game? How might this change affect your strategies? Write a brief summary explaining your reasoning.

Teacher Talk

Where's the Mathematics?

This game provides an exciting way for students to explore strategies based on probability. Students may initially see *Freeze Before Fifty* as a game of chance since spinners and dice are used to change the number of Snap Cubes collected. However, choosing when to freeze gives players a degree of control over the game, so some skill is involved.

There is no advantage in going first since the effect of rolls on the Snap Cube collection is left to chance. Either player can freeze on any turn as long as that player thinks he or she is close enough to 50. The player with the most cubes at the end of any turn has a slight advantage because this player has the option of freezing play, an option the other player does not have until it is his or her turn. However, this situation could be reversed on the next turn depending on the spin of the spinner and roll of the number cube.

Strategies that help students decide when to freeze are useful. You may hear some of the following strategies suggested:

"If you have at least 8 cubes more than the other player, freeze. The other player will have a hard time getting more than you without going over 50."

"If the game is really close, wait until you have 30 to 35 cubes before you freeze. The other player will probably go over trying to beat you."

"Always freeze if you have more than 40 cubes. A few more cubes aren't worth risking going over."

In *Freeze Before Fifty*, students need to assess risk on each turn. Students may do this by figuring out the probability that a particular situation will occur, or they may go by what feels lucky. Many students, and even adults, are convinced that luck gives them the special ability to roll certain numbers on a die or control the results of a spin. If students play only one or two games, this perception may be reinforced. You may want to have students play many games over a period of time to get a clearer sense of what is likely to occur.

Students may have different ways of comparing the probabilities of adding 4 or multiplying by 2. The actual probabilities are found by multiplying the chance of spinning a particular operation by the chance of rolling a particular number. The probability of adding 4 is $2/3 \times 1/6$, which equals 2/18 or 1/9. The probability of multiplying by 2 is $1/3 \times 1/2$, which equals 1/6. So the chance of multiplying by 2 is greater than the chance of adding 4.

For both versions of the game, there is a 2/3 chance of spinning the addition operation and a 1/3 chance of spinning multiplication. When this information is combined with chances of rolling various numbers, students can calculate the chance of their going over 50 on a turn. This table shows the six different possible outcomes of rolling and spinning for the first version (three 2s, two 3s, and one 4).

Version 1				
Number Rolled	**Probability**	**Operations Spin**	**Probability**	**Combined Probability**
2	3/6, or 1/2	Addition	4/6, or 2/3	2/6, or 1/3
2	3/6, or 1/2	Multiplication	2/6, or 1/3	1/6
3	2/6, or 1/3	Addition	4/6, or 2/3	2/9
3	2/6, or 1/3	Multiplication	2/6, or 1/3	1/9
4	1/6	Addition	4/6, or 2/3	2/18, or 1/9
4	1/6	Multiplication	2/6, or 1/3	1/18

Note: The total combined probability is 18/18, or 1.

In the second version of the game, the strategies change somewhat, since there is a greater range of numbers that can be rolled. In the first game, the range is from 2 to 4. In the second game, the range is from 3 to 10. Because of the greater range, students may play more cautiously and freeze sooner. Not only does it make it more likely that a player will go over 50 if multiplication is spun, but it also gives a player starting over a greater chance of scoring close to 50 in only one or two turns. This means that players may be more willing to go over 50 and start again.

Students may mistakenly assume there are only 12 possible dice rolls in the second version since there are 6 faces on each die. To find the distribution of possible outcomes for both dice, students must consider the possible ways a particular sum can be made using both dice. On the number cube, some numbers are repeated. Each of these faces of the number cube must be treated as different from its other faces with the same number. The possible outcomes for the two dice used in the second version are shown in the graph below. For each combination, the result for the number cube marked 2, 2, 2, 3, 3, and 4 is always listed first. Lowercase letters are used to distinguish the six number cube faces. These numbers are then added to the number from the regular die roll.

			2c + 3	2c + 4	2c + 5	2c + 6		
		2c + 2	2b + 3	2b + 4	2b + 5	2b + 6		
		2b + 2	2a + 3	2a + 4	2a + 5	2a + 6		
2c + 1	2a + 2	3b + 2	3b + 3	3b + 4	3b + 5	3b + 6	
2b + 1	3b + 1	3a + 2	3a + 3	3a + 4	3a + 5	3a + 6	
2a + 1	3a + 1	4 + 1	4 + 2	4 + 3	4 + 4	4 + 5	4 + 6
3	4	5	6	7	8	9	10

HALF CHANCE

- Chance
- Spatial visualization
- Game strategies

Getting Ready

What You'll Need

Cuisenaire Rods, 2 sets per pair
Half Chance Game board, page 116
Half Chance Spinner A, page 117
Half Chance Spinner B, page 118
Activity Master, page 109

Overview

In this activity, two players use spinners to determine the color and number of Cuisenaire Rods to place on a rectangular grid in an effort to cover one-half the grid. In this activity, students have the opportunity to:

- develop strategies based on probability and spatial reasoning
- enhance number sense about fractions
- strengthen strategic thinking skills

Other *Super Source* activities that explore these and related concepts are:

Freeze before Fifty, page 66

Block Path, page 74

The Activity

On Their Own (Part 1)

Half Chance is a game for 2 players. The object is to be the first player to cover half of the game board with Cuisenaire Rods. Can you find the winning strategies?

- Choose which side of the Half Chance game board you will play and then spin Spinner A. The player who spins the higher number goes first.

- On your turn, you can either spin to add Cuisenaire Rods to the board or spin to remove them. You must decide before you spin.

- Adding Cuisenaire Rods: Decide whether to use Spinner A, Spinner B, or a combination of the 2 spinners and then spin twice. The outcome of the first spin tells which color Cuisenaire Rod to add to the board. The outcome of the second spin tells how many of that color you may add to the board. For example, if you spin light green on Spinner A and 2 on Spinner B, you add 2 light green rods.

- Fraction Bonus Turn: If at the end of your turn, you have covered exactly 1/4 (25 squares), 1/2 (50 squares), or 3/4 (75 squares) of your portion of the game board, you take an additional turn.

- *Placement of Cuisenaire Rods:* Once a rod is played, it cannot be moved unless it is removed from the board during a player's turn. Added rods may be played in any empty space. If you cannot add all the rods indicated by your spin, you add no new rods and instead remove one rod of any color from the board.
- *Out of a Color:* If you do not have enough Cuisenaire Rods of the color spun to make your play, spin both spinners again.
- Play continues until 1 player covers his or her half of the game board.
- Play several games of Half Chance. Be ready to discuss winning strategies.

Thinking and Sharing

Invite pairs to talk about their game and describe some of the strategies they found.

Use prompts like these to promote class discussion:

- Is *Half Chance* a game of chance, of skill, or of both? Explain.
- How did you decide which spinners to use on a turn?
- How did you decide where to place the Cuisenaire Rods on the grid?
- Did you place any Cuisenaire Rods that you had to go back and remove? Explain.
- Did you ever get a Fraction Bonus turn? Was it by chance? Did you use any strategies to achieve a certain fraction?
- What strategies did you try as you played the game? Did your strategies work? What would you do differently if you were to play *Half Chance* again?

On Their Own (Part 2)

What if... you change the game of Half Chance so that added Cuisenaire Rods must connect? How would this change your game strategies?

- Play the game again, except this time any rods that you add must connect along a side to form a block or a chain.

  Okay: Not okay:

- Be ready to explain how the change in the rules affected your game strategies.

Thinking and Sharing

Invite students to compare the two versions of the game.

Use prompts like these to promote class discussion:

- Which version of the game was more fun? Why?
- Which version required more skill?
- What game strategies did you use to play this version of *Half Chance*?
- How did the rule change affect your game strategies?

Write a letter to a new student who will be playing *Half Chance* for the first time. Give the student advice on the best strategies to use. Include your strategies for choosing which spinners to use, placing rods, and earning Fraction Bonus turns.

Teacher Talk

Where's the Mathematics?

This game provides an engaging way for students to explore fraction concepts while developing strategies based on probability and spatial reasoning. At first, students may see this activity as a game of chance because spinners are used to decide how many rods may be placed. However, choosing which spinners to spin and where to place rods gives players a great deal of control over the game. *Half Chance* is a game of both chance and skill.

Initially students may not understand the strategy behind choosing which spinner to use. As they examine the spinners and think about the size of the Cuisenaire Rods, they may realize that selecting Spinner A maximizes the choice of length and number of rods they may use and therefore the amount of area covered. If Spinner A is spun twice, most turns will cover an area of 10 cm² or more, compared to 3 cm² or more for Spinner B and 6 cm² or more for a combination of A and B. While spinning Spinner A twice is a strategy that works well at the beginning of the game, it does not work well toward the end when there are only small spaces left to be covered.

Students may calculate the theoretical probability of covering a certain number of square centimeters for each two-spinner combination as an aid in choosing spinners for game situations. The number of possible spins resulting from any two-spinner combination is found by multiplying the possible number of spins from one spinner by the possible number from the other spinner. To find theoretical probability, students write a ratio of the number of possible ways to spin a certain area to the total number of possible spins.

Spinners A and A		Spinners A and B (either order)		Spinners B and B	
Area	Probability	Area	Probability	Area	Probability
1 cm²	1/36	1 cm²	1/18	1 cm²	1/9
2 cm²	1/18	2 cm²	1/9	2 cm²	2/9
3 cm²	1/18	3 cm²	1/9	3 cm²	2/9
4 cm²	1/12	4 cm²	1/9	4 cm²	1/9
5 cm²	1/18	5 cm²	1/18	6 cm²	2/9
6 cm²	1/9	6 cm²	1/6	9 cm²	1/9
8 cm²	1/18	8 cm²	1/18		
9 cm²	1/36	9 cm²	1/18		
10 cm²	1/18	10 cm²	1/18		
12 cm²	1/9	12 cm²	1/9		
15 cm²	1/18	15 cm²	1/18		
16 cm²	1/36	18 cm²	1/18		
18 cm²	1/18				
20 cm²	1/18				
24 cm²	1/18				
25 cm²	1/36				
30 cm²	1/18				
36 cm²	1/36				

In the second version of the game, students may find that playing a long, winding chain of rods can create problems on a later turn. However, if this is the only way to make a play, students may find it a desirable choice; particularly if the other player has covered substantially more area.

Both plays cover 12 cm², but the more compact play on the right makes later moves easier.

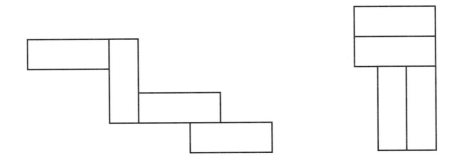

Fraction Bonus turns can affect the outcome of a game. Gaining an extra turn to cover more area makes it easier to fill a particular space and can give a player an obvious advantage. At first, students may ignore the Fraction Bonus, achieving it by chance alone. As they see the advantage, they may play to gain the extra turn. Students may decide to keep a tally of the area covered to make it easier to know when they have gained the bonus. Others may block off the board into 1/4, 1/2, and 3/4 sections and then attempt to cover those sections exactly as they play.

BLOCK PATH

- Chance
- Measurement
- Spatial visualization

Getting Ready

What You'll Need

Pattern Blocks, approximately half of 1 set per pair

Dice, 2 per pair

Adding machine tape, or long strip of paper, 1 per player

Meter stick, 1 per pair

Activity Master, page 110

Overview

Students play a game in which they use Pattern Blocks, selected by chance and skill, to build a path that ends closest to the finish line. In this activity, students have the opportunity to:

- develop strategies based on probability and spatial reasoning
- make estimates involving length
- develop strategic thinking skills

Other *Super Source* activities that explore these and related concepts are:

Freeze Before Fifty, page 66

Half Chance, page 70

The Activity

On Their Own (Part 1)

> **Block Path is a game in which players build paths of Pattern Blocks. The object of the game is to build a path that ends closest to the finish line without going over it. Can you find the winning strategies?**
>
> - Work with a partner. Each of you needs a paper track that is about 70 centimeters in length. Mark a starting line at one end of the paper. Mark a finishing line 50 centimeters away from start. All moves must fit inside this strip of paper.
> - Roll a die and collect that number of one kind of Pattern Block shape.
> - To begin your path, place your first block at the starting line. Then continue to place blocks so that each new block touches the last block you placed. Each block must be closer to the finish line than the previous block.
>
>
>
> Okay   Not Okay   Okay
>
>
>
> - After your partner has taken a turn and added to his or her own path, roll again and take that number of a *different* Pattern Block shape. Add them to your path without rearranging the blocks that you placed on your previous turn(s).

- Continue playing until each of you has had six turns, and you have used each kind of Pattern Block shape.
- The player whose last piece is closest to the finish line without going over it is the winner. If both of you go over the finish line then the player whose last piece is closest to the finish line is the winner.
- For each player, record the total distance (in centimeters) from the starting point to the end of the path and the total number of each shape used by each player.
- Play several games of Block Path.
- Be ready to talk about your strategies.

Thinking and Sharing

Have pairs discuss the strategies they used while playing the game.

Use prompts like these to promote class discussion:

- How did you decide which Pattern Block shape to choose for each die roll?
- If you roll a small number on your first turn, which kind of block would you choose? Why?
- What strategies did you use if you wanted to cover a lot of distance? a little distance?
- Did you use the same strategy on each of your turns or did your strategy change during the game? Explain.
- Did you use the same strategy in every game? Explain.

On Their Own (Part 2)

What if... you use two dice instead of one, spell out the word for the number rolled, and use that many Pattern Blocks to make your path? How will this change your game strategies?

- Start playing Block Path again, but this time roll a pair of dice. For each roll, add the numbers on the dice and then figure out the number of letters in the number you rolled. For example, for a sum of *ten*, you collect 3 Pattern Blocks because there are 3 letters in the word *ten*. For a sum of *eleven*, you collect 6.
- Collect and place blocks onto a pathway as you did in Part 1.
- The winner is determined in the same way as before. The game is over after each of you has had 6 turns.
- As before, record the total length of the paths in centimeters and the total number of each shape used by each player.
- Play several games of this version of Block Path.
- Be ready to discuss how the change in the rules affected your game strategy.

Thinking and Sharing

Have pairs share their game records and any changes in their strategies. Create a chart with the following headings: *number rolled, number word, number of letters, probability of dice rolls, combined probability.*

Use prompts like these to promote class discussion:

- How did the rule change affect your game strategies?
- What was the smallest number of blocks you could place? The largest number?
- Based on the letters in the number words, which number of blocks did you place most often? least often?
- Which version of the game did you like better? Why?
- Do you think *Block Path* is more a game of chance or of skill? Explain.
- If you were to give someone advice about how to win at this game, what would you say?

Games that are played with dice usually involve chance, but many, such as *Block Path*, also have an element of skill. Think of a game that you have played that requires dice. Write a summary explaining whether the game is more a game of chance or a game of skill. Use specific examples from the game to prove your point.

Teacher Talk

Where's the Mathematics?

In this activity, students develop strategies based on chance and spatial reasoning as they attempt to combine Pattern Blocks to build a path that is as close as possible to 50 centimeters in length. As students assign die rolls to various shapes, they may take into consideration which shapes work together easily to best fill the space they have left to cover.

To maximize length, students will learn to use the longer diagonal of the tan rhombus, the diagonal of the hexagon, and the longer base of the trapezoid.

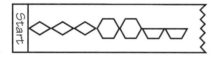

To minimize length and fill space, students may use triangles with the sides touching.

In either version of *Block Path*, students' strategies generally change as they get closer to the end of the game. Since students cannot control which number comes up next, they will generally use their first few rolls to cover as much distance as possible. Hexagons and trapezoids are useful in the beginning moves because they cover a lot of length. It may be difficult to place these pieces in the later moves of the game, because of their large size. It is generally wise to save the triangles until the end of the game since they can be placed either with their points touching to cover distance, or with their sides touching to minimize distance. For the same reasons, some students

may also find tan rhombuses to be a good choice for one of the final turns of the game.
Your students' records of the first version of *Block Path* may look something like this:

Players	1	2	3	4
Total Length	49.5 cm	48.5 cm	52.3 cm	50.5 cm
Green Triangle	2	2	2	5
Orange Square	1	6	5	4
Blue Rhombus	2	4	4	5
Tan Rhombus	5	3	2	2
Red Trapezoid	3	2	5	4
Yellow Hexagon	4	1	2	3

There is an element of chance in this game since students cannot control which number will come up next on the die. In the first version, the chance of rolling any number is 1 in 6. In other words, the chances of rolling any number on one die are equal. Students can investigate this idea further by making a combined frequency graph of their die rolls from their game records.

In the second version of the game, the chances of rolling the various numbers are not equal. Although the total of the dice may be any number from 2 through 12, the letters in the number words range only from 3 to 6.

The theoretical probabilities of placing 3, 4, 5, or 6 blocks (based on the letters in the number words) can be figured by first determining the number of possible ways to roll each total from 2 through 12. If students have had prior experience with probability and dice, they may already realize that there are 36 possibilities. The frequency graph below shows the distribution of the possible rolls of two dice.

```
                        X
                    X   X   X
                X   X   X   X   X
            X   X   X   X   X   X   X
        X   X   X   X   X   X   X   X   X
    X   X   X   X   X   X   X   X   X   X   X
    2   3   4   5   6   7   8   9   10  11  12
```

Three of the number words—*two, six,* and *ten*—have 3 letters. By adding the chances of getting each of these numbers, students can find the probability of rolling a number that results in the placement of 3 Pattern Blocks. The chance of rolling a 2 is 1/36, the chance of rolling a 6 is 5/36, and the chance of rolling a 10 is 3/36 for a total of 9/36, or 1/4. In other words, there is a 1 in 4 chance of a dice roll resulting in the placement of 3 Pattern Blocks.

The probabilities for each possibility are shown below:

Number of Blocks Placed	Possible Dice Rolls	Probabilities	Combined Probability
3	two, six, ten	1/36, 5/36, 3/36	9/36 or 1/4
4	four, five, nine	3/36, 4/36, 4/36	11/36
5	three, seven, eight	2/36, 6/36, 5/36	13/36
6	eleven, twelve	2/36, 1/36	3/36 or 1/12

From game experiences and calculations similar to those above, students may see that there is greater than a 1 in 3 chance that they will be required to place five Pattern Blocks in any given turn. This observation may affect game strategies as students seek to minimize the length added by the blocks. Students may decide to add any length they may need in later turns instead of during the opening turns of the game.

Both game versions require skill in spatial reasoning and planning. An understanding of probability can help students make skillful moves in either game. However, the element of chance is great enough that even the most skillful players may find themselves far beyond the finish line because of an unfortunate dice roll.

Investigating Permutations

1. Mall Madness, page 80   (Color Tiles or Snap Cubes)
2. Quilt Squares, page 85   (Tangrams)
3. Starting Five, page 90   (Snap Cubes)

This cluster contains activities designed to solidify students' understanding of permutations. The first two lessons in the cluster also include the use of combinations. This material could be presented as an introduction to or reinforcement of ideas and methods pertaining to permutations.

1. Mall Madness (Exploring combinations and permutations)

This activity can be used as an introduction to combinations and permutations. *On Their Own* Part 1 gives informal definitions of each term and explains how to distinguish between the two concepts. The sequence of this activity gives students the opportunity to understand how combinations can lead to permutations.

In *On Their Own* Part 1 and Part 2 students are expected to organize their data regarding color combinations and arrangements. This allows students to explore the various ways of tracking different combinations and permutations, and it gives them the opportunity to figure out which way (chart, list, diagram) works best for them.

Where's the Mathematics? (page 83) gives the teacher suggestions on how to guide students to explore the formal calculations for permutations using factorials. This lesson suggests using Color Tiles, but students could use Snap Cubes as a substitute.

2. Quilt Squares (Extending combinations and permutations)

In this activity, students use Tangrams to create various colored patterns on a *Quilt Squares Pattern* page. Teachers may want to review the mathematical means for finding the number of possible arrangements.

In *On Their Own* Part 2 students are asked to predict the number of color arrangements possible using up to four different colors. Students may make predictions based on their work from Part 1 in which they found the arrangements using up to two different colors.

Where's the Mathematics? on page 89, gives an explanation about the formal calculations involving circular permutations.

3. Starting Five (Investigating permutations)

Teachers might want to review combinations and permutations before starting this activity. Students are given a hypothetical situation and are then expected to simulate the situation using Snap Cubes. Students are exposed to the Basic Counting Principle as they explore patterns in their data and they gain experience in using objects to model a situation.

In *On Their Own* Part 1 students are expected to record data and they should have had previous experience with different strategies for arranging and finding permutations.

Where's the Mathematics? (page 92) gives explanations regarding factorials, and patterns. There are also approaches to finding all of the possible arrangements from Parts 1 and 2.

MALL MADNESS

- Permutations
- Combinations
- Organizing data

Getting Ready

What You'll Need

Color Tiles, 12 each of 4 colors per pair

Activity Master, page 111

Overview

Students search to find all possible three-color combinations and permutations using the four different colors of Color Tiles. In this activity, students have the opportunity to:

- explore similarities and differences in combinations and permutations
- organize and record solutions using lists, charts, and diagrams

Other *Super Source* activities that explore these and related concepts are:

Quilt Squares, page 85

Starting Five, page 90

The Activity

On Their Own (Part 1)

> Dominic is an architect who plans to have each level of a shopping mall painted red, blue, green, or yellow. The mall has three levels. Dominic wants each level to be a different color. How many possible color combinations are there for the three levels? How many different ways can the colors be arranged?
>
> - Working with a partner, use Color Tiles to represent the different levels of the shopping mall. Find all the possible color combinations for the three levels. Order doesn't matter in a combination. *Red-blue-green is the same as green-blue-red.*
> - Make a list of the combinations. Make sure your list does not contain repeats of the same combinations.
> - Now find all the ways the colors can be arranged. These are called permutations. Order does matter in a permutation. *Red-blue-green is therefore not the same as green-blue-red.*
> - Record each permutation. Find a way to organize your solutions.
> - Compare the numbers of combinations and permutations. Be ready to discuss their relationship.

Thinking and Sharing

Have the students work together to make a class list of possible combinations and permutations.

Use prompts like these to promote class discussion:

- How did you make sure that you had found all the possible combinations of colors?
- How did you make sure that you had found all the possible permutations?
- Did you organize your work in any special way? Explain.
- What strategies did you use to make sure you were not repeating a combination?
- Compare the numbers of combinations and permutations. How are these numbers related? Explain.
- Based on your class data, what formula would you create to predict or calculate the number of 3-color permutations that could be made from 4 colors?

On Their Own (Part 2)

What if... the levels of the mall do not have to be different colors? How many color combinations are there? How many ways can the colors be arranged?

- Working with your partner, use Color Tiles to find the possible 1-color, 2-color, and 3-color combinations. *Red-red-green* and *green-green-red* would both be considered a single 2-color combination of *red-green*.
- Find ways to record your solutions.
- Now use the Color Tiles to find all the possible permutations. Find ways to record your work.
- Be ready to explain the relationships between the number of combinations and permutations.

Thinking and Sharing

Have the students work together to make a class list of possible combinations. Invite pairs to display their permutations.

Use prompts like these to promote class discussion:

- What affect did the possibility of same-colored levels have on the possible numbers of combinations and permutations?
- How did you make sure that you had found all the possible combinations and permutations of colors?
- Did your work from the first activity help you in any way? Explain.
- What strategies did you use to record your solutions?
- How did you make sure you were not repeating a combination or permutation?
- Were there any patterns that could help you predict or calculate the number of combinations or permutations that could be made from any number of colors? Explain.

For Their Portfolio

Five students have worked together to perform a science experiment. The teacher will choose two students from the group to present the experiment to the class. Write a summary explaining the difference between a list of combinations and a list of permutations. Which list do you think would be more helpful to the teacher in this situation? How many possible combinations and permutations would the teacher have to choose from in order to select two students from the group? Include a list and/or any charts or diagrams that might be helpful for the teacher to see every possibility.

Teacher Talk

Where's the Mathematics?

In this activity, students explore the difference between combinations and permutations. Students also have the opportunity to use an organized approach to find and check possible solutions.

In both activities, students choose three of four colors to paint the three levels of a shopping mall. In the first activity, students must choose three different colors. In the second activity, this limitation is removed. Students may be surprised at how many more solutions are possible once all combinations of colors are allowed.

The number of possible combinations in the first activity is 4. After making the first combination, students may find the remaining combinations simply by trading Color Tiles. For example, after finding the combination *red-yellow-green*, students may trade the green tile for a blue one to make a new combination. Students who have had experience finding permutations may have difficulty ignoring the order of the colors while looking for new combinations. The possible combinations are *red-yellow-green, red-yellow-blue, red-blue-green,* and *blue-yellow-green*.

There are 24 possible permutations using three different colors. Students may use a variety of methods to organize their search for solutions. Students may make random arrangements and then check to see whether the order is a new one. They may attempt to make all the solutions that have a certain color in the first position, varying the final colors in some organized fashion. They may also make lists and diagrams in addition to their drawn chart of the solutions.

One way to organize the solutions is with a tree diagram. The tree diagram below shows the solutions for the permutations beginning with red.

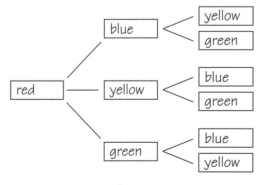

Students may also use the list of combinations to generate the possible permutations. For example, from the combination *red-yellow-green*, students can, through exploration and manipulation of the order, quickly find six possible permutations. This approach will help students see that a combination describes the contents of a set while a permutation describes both the contents and the arrangement of the items in the set.

Combination	Permutations	
red-yellow-green	red-yellow-green	yellow-green-red
	red-green-yellow	green-red-yellow
	yellow-red-green	green-yellow-red

Students may also use logical reasoning to find the number of possibilities. For example, using the tree diagram they may reason that since there are 6 solutions that begin with red and there are 4 colors that can occupy the first position in the chart, there must be 6 × 4, or 24 possible solutions. Using the chart above, they may see that there are 6 permutations for each of the 4 combinations, again resulting in 24 solutions.

Another way to find the number of possible permutations is by writing an equation. Students may reason that there are 4 possible color choices for the first level of the shopping mall. Since the levels must be different colors, there are 3 possible color choices for the second level and 2 possible choices for the third level. Therefore, there must be 4 × 3 × 2, or 24 permutations.

As students explore the definitions and relationships between combinations and permutations, they may make statements like "There are always more permutations than there are combinations." They might also say, "You can use the combinations to help you think of all the permutations," and "Every combination is a permutation, but different permutations are not always different combinations."

This exploration helps lay the foundation for later formal calculations involving the use of factorials. To find the number of permutations of n objects taken r at a time, use the formula below. The variable n represents the number of colors and r represents the number of floors in the mall.

$$n\,P\,r = \frac{n!}{(n-r)!} \text{ which, in this case, would be } \frac{4!}{(4-3)!} = \frac{4 \times 3 \times 2 \times 1}{1} = 24$$

The number of combinations is found using the following formula for n objects taken r at a time:

$$\frac{n!}{r!\,(n-r)!} = \frac{4!}{3!\,(4-3)!} = \frac{4 \times 3 \times 2 \times 1}{3 \times 2 \times 1(1)} = \frac{24}{6} = 4$$

In the second activity, the students are first asked to find the number of possible 1-color, 2-color, and 3-color combinations. They should be concerned with what colors are used, not where they are used or how often they are used. As long as the colors can be used more than once, there are 14 possible color combinations. Students should realize that they have already found the 3-color combinations. The 1- and 2-color combinations are:

Colors used in the 1-color combinations	Colors used in the 2-color combinations	
red	red-blue	blue-green
blue	red-green	blue-yellow
green	red-yellow	green-yellow
yellow		

Students may recognize that the solutions for both combinations and permutations for the first activity are part of the larger set of solutions for the second activity. From this, they know the number of possible permutations must be greater than 24. Once again, they can build the arrangements using the list of combinations. They can also use the number of combinations to reason the number of permutations.

They may reason that for every 1-color combination there is one permutation, and for every 2-color combination there are six possible permutations. For example, the combination *red-blue* can be used to make the permutations *red-red-blue, red-blue-red, blue-red-red, blue-blue-red, blue-red-blue,* and *red-blue-blue*. They already know there are four 3-color combinations, each yielding 6 permutations. With this knowledge, they can calculate the total permutations as follows:

Four 1-color combinations with 1 permutation each make 4×1, or 4 permutations.
Six 2-color combinations with 6 permutations each make 6×6, or 36 permutations.
Four 3-color combinations with 6 permutations each make 4×6, or 24 permutations.

$4 + 36 + 24 = 64$ total permutations

Students can also calculate the possible arrangements by multiplying. If any of the four colors can be used to paint any of the three levels, there are four possible colors for each level for a total of $4 \times 4 \times 4$, or 64 permutations.

QUILT SQUARES

- Combinations
- Permutations
- Organizing data

Getting Ready

What You'll Need

Tangrams, 16 small triangles, 4 of each color, per group

Quilt Squares Patterns, page 119

Quilt Squares Record, page 120

Colored pencils or markers (optional)

Activity Master, page 112

Overview

Students work in small groups using Tangram pieces to find all the possible arrangements of colors that can be used to fill a quilt square pattern. In this activity, students have the opportunity to:

- work informally with combinations and permutations
- organize and record solutions
- explore how rules are used to limit the contents of a set
- make predictions

Other *Super Source* activities that explore these and related concepts are:

Mall Madness, page 80

Starting Five, page 90

The Activity

On Their Own (Part 1)

Quilt squares are made from cloth geometric shapes that are pieced together. Once a pattern is chosen, the quilt maker sometimes varies the colors of the shape to create an interesting design. Can you find all the possible arrangements of color that can be used to fill a quilt square?

- Work in a small group. Each of you needs a copy of the *Quilt Squares Patterns* page. The quilt maker plans to use white fabric for the center square. The triangle pieces that border the square can be red, blue, yellow, or green. But there can be no more than two different colors of triangles in any quilt square.

- Use your small Tangram triangles to find all the possible arrangements of two colors. Check to make sure you have not created any repeats. Turning a quilt square does not count as a different quilt square.

Example:

These should be considered one quilt square.

- Record all the possible quilt squares on your *Quilt Squares Record* page by coloring or labeling the triangles.
- Count the total number of different arrangements you have found. Be ready to share all the possible quilt squares that can be made with two of the four possible colors.

Thinking and Sharing

Invite groups to display their work. Based on their findings, create a list of the possible arrangements.

Use prompts like these to promote class discussion:

- How many different arrangements did you find?
- How did you make sure that you had found all the possible arrangements?
- Did you organize your work in any special way? Explain.
- What strategies did you use to make sure you were not repeating a solution?
- Do you see any way to predict or calculate the number of quilt squares that can be made?

On Their Own (Part 2)

What if... you can use up to four colors of triangles in one quilt square? How many color arrangements are there? Can you find them all?

- Make a prediction as to how this change in the rules will affect your work. How many different quilt squares do you think there will be using up to four colors of small triangles? Record your prediction.
- Work with your group to find all the possible quilt squares. Use the small Tangram triangles and the *Quilt Squares Patterns* page.
- Remember, you can use up to four colors in each quilt square. Each arrangement of colors must be different. Use the *Quilt Squares Record* page to record your solutions.
- Be ready to explain how you know that you have found all the possible solutions.

Thinking and Sharing

Have groups share their predictions and display their work.

Use prompts like these to promote class discussion:

- How did you make your predictions? Did you use any mathematical procedures? If so, what did you do?
- What strategies did you use to find the possible solutions?
- How did you organize your work?
- What strategies did you use to make sure each quilt square was different?
- Did your work for the first activity help you in any way? How did you use it?

For Their Portfolio

> A circular stained-glass window has a border of panes made from 5 colors. How would you go about finding the number of possible arrangements that you can make using all 5 colors. Turning the border does not count as a different arrangement. Write a summary explaining how you would go about finding the number of possible arrangements there would be for 5 colors.

Teacher Talk

Where's the Mathematics?

In this activity, students search for different arrangements of colors using Tangram pieces. As students work, they have the opportunity to use an organized approach to finding and checking solutions. Students may also explore mathematical means for calculating the number of possible quilt squares.

In both activities, students choose from 4 colors of small Tangram triangles to complete quilt squares. In the first activity, the students can use no more than 2 colors at a time. In the second activity, students can use up to 4 colors. In both activities, rotating a quilt square does not count as a new solution.

Using up to 2 different colors, there are 28 possible solutions. As students work to find these solutions, they may combine colors randomly or they may use an organized approach. Some students may form one combination and then exchange Tangram pieces in a systematic fashion to come up with new solutions. Students may prefer to make organized lists or charts to keep track of the possible combinations.

One approach your students may use is shown on the next page. This chart breaks the task down into three possible kinds of combinations: 4 triangles of the same color; 3 triangles of one color and 1 of another color; 2 triangles of one color and 2 of another color. Students may experiment to find the number of arrangements possible for each combination.

Combination Type	Logical Reasoning	Number of quilt squares
4 of the same color	Since there are 4 colors, there are 4 one-color combinations.	4
3 of one color and 1 of another color	There are 3 two-color combinations that could have 3 red triangles and one of another color: *red-blue* *red-green* *red-yellow* Therefore, there must be 3 combinations each that could have 3 green, blue, or yellow triangles. *3 combinations × 4 colors = 12 quilt squares*	12
2 of one color and 2 of another color	There are six possible 2-color combinations: *red-blue, red-green, red-yellow, blue-green, blue-yellow,* and *green-yellow*. For each 2-color combination, there is 2 different arrangements of colors: *6 combinations × 2 arrangements = 12 quilt squares*	12
	Total: 28 different quilt squares	

As students compare quilt squares to find repeated solutions, they realize the importance of using an organized approach to finding solutions. Students may use charts, lists, or tree diagrams in addition to the *Quilt Squares Record* page.

In the second activity, there are 70 possible solutions. In making their predictions, students may realize that the solutions from the first activity are also solutions for the second. Some students may assume that since the number of colors that are used has doubled, the number of solutions from the first activity should be doubled for a prediction of 56 quilt squares. If students have had experience with permutations, they may make a much higher prediction. They may mistakenly suppose that using all 4 colors will result in 4!, which equals 4 x 3 x 2 x 1, or 24 possible arrangements. This would be the case if rotated squares were considered different arrangements.

To find the number of possible arrangements using up to 4 colors, students may use the following reasoning:

Combination Type	Logical Reasoning	Number of quilt squares
Using up to 2 colors	Repeat the results from the first activity.	28

3-color combinations	There are four possible 3-color combinations. They are:
	
	    red-blue-green
	    red-blue-yellow
	    red-green-yellow
	    blue-green-yellow
	
	In each arrangement, some color must be used twice. Letter A represents the color used twice, Letters B and C represent the remaining colors.
	
	For each of the 3-color combinations, each color can be used in the place of A. Therefore, there are 9 arrangements for each combination (3 colors × 3 arrangements) and a total of 9 × 4, or 36 arrangements for quilt squares with 3 colors.
		36

4-color combinations	Every arrangement must include every color, therefore find the permutations that begin with a certain color. All other possible arrangements are rotations of one of these arrangements:
	
	    red-blue-yellow-green
	    red-blue-green-yellow
	    red-yellow-blue-green
	    red-yellow-green-blue
	    red-green-blue-yellow
	    red-green yellow-blue
		6

Total: 70 different quilt squares

This exploration lays the foundation for formal calculations involving circular permutations. Since rotations are not considered solutions, students may notice that the arrangements in this activity are circular. In fact, the number of arrangements for 4 colors can be calculated by writing an equation. The number of permutations of n distinct objects (in this case, colors) that can be arranged in a circle is $(n-1)!$. Since there are 4 colors, the number of circular permutations is $(4-1)!$ or $3!$, which equals $3 \times 2 \times 1$, or 6.

STARTING FIVE

- Permutations
- Looking for patterns
- Making predictions
- Logical reasoning

Getting Ready

What You'll Need

Snap Cubes, 10 each of 7 different colors per pair

Activity Master, page 113

Overview

Students use Snap Cubes to represent players on a basketball team. Students need to arrange the players in as many different ways as possible to model various situations. In this activity, students have the opportunity to:

- enhance their understanding of permutations and factorials
- discover patterns in data
- use patterns to make predictions

Other *Super Source* activities that explore these and related concepts are:

Mall Madness, page 80

Quilt Squares, page 85

The Activity

On Their Own (Part 1)

> The Long Tall basketball team has five star players. At every game, the players want to run out onto the floor in a different order. How many possible orders, or arrangements, of players are there? Can you discover a pattern that will help you figure out how many ways there are to arrange any number of players?
>
> - Working with a partner, use 5 different Snap Cube colors to represent the five basketball players.
> - Find all the possible ways to arrange the five players as they run out onto the court one at a time.
> - Find ways to record your solutions, first giving the number of possible arrangements for 1 player, 2 players, 3 players, 4 players, and 5 players. Make sure you do not repeat the same arrangement of players.
> - Look for a pattern in your data that would help you predict the number of possible arrangements of any number of players. Predict the number of different arrangements there are for 10 players.
> - Be ready to discuss the number of arrangements for each number of players.

Thinking and Sharing

Invite pairs to give solutions for each of the arrangements from one through five players. Create a class chart indicating the number of different arrangements for each number of players.

Use prompts like these to promote class discussion:

- How did you make sure that you had found all the possible arrangements of the colors?
- As you worked, did you use strategies to avoid repeating any arrangements?
- Do you see any patterns in the data? If so, explain.
- How many different arrangements of ten players did you predict? Justify your answer.
- Can you use the patterns you discovered to predict possible arrangements of any number of players?

On Their Own (Part 2)

What if... there are seven star players on the team, but one player has become so well-liked by the fans that this player always runs out onto the floor either first or last? How many possible arrangements of the players are there now?

- Use different colors of Snap Cubes to represent the players.
- Find all the possible ways to arrange seven players if one player has to be either first or last every time.
- Find ways to record your work. Make sure you don't repeat the same arrangement of players.
- Is there a way to find the answer without making and recording every combination? Try it. Look for patterns.
- Be ready to explain how you got your answer.

Thinking and Sharing

Have pairs share their answers and the processes they used to find them.

Use prompts like these to promote class discussion:

- How many different arrangements did you find?
- How did you find your solution? What did you do about the player that had to run out either first or last?
- Did your work from the first activity help you in any way? If so, how?
- How did you organize your work so that you could look for patterns?
- How can you be sure that you found every possible arrangement of players?
- How did you make sure that you were not repeating any arrangements?

For Their Portfolio

The team decides to have a photo taken of its five starters, but the two guards refuse to stand next to each other. Suppose you need to find the number of possible arrangements of players. How would you do it? Write a brief summary explaining how you would organize your work and what calculations you would perform to find an answer.

Teacher Talk

Where's the Mathematics?

In this activity, students are exposed to the Basic Counting Principle, a concept that students will encounter in more detail in their study of permutations and combinations. As students explore patterns in their data, they lay the foundation for a greater understanding of how factorials are used to solve problems. Students also gain experience in using objects to model a situation.

Students may use different strategies for finding the different arrangements. Some may search randomly for new arrangements while others may use a more systematic approach. Students may use diagrams or tables to organize their arrangements.

A tree diagram showing the six possibilities for three colors (red, blue, and yellow) is shown below.

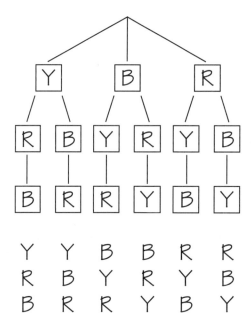

As the number of colors increase, students are likely to discover that they need some kind of strategy in order to avoid repeating arrangements. One approach is to find all the arrangements that start with a particular color or a combination of colors. The following chart lists the possible 5-color arrangements that begin with red. In each column, red is paired with a different possible second color. The colors represented are red, blue, yellow, green, and white.

RB	RY	RG	RW
RB Y GW	RY B GW	RG B YW	RW B GY
RB Y WG	RY B WG	RG B WY	RW B YG
RB G YW	RY G BW	RG Y BW	RW G BY
RB G WY	RY G WB	RG Y WB	RW G YB
RB W YG	RY W BG	RG W BY	RW Y BG
RB W GY	RY W GB	RG W YB	RW Y GB

For five or more colors, listing all the arrangements is tedious. Logical reasoning can make the task easier. For example, students looking for the possible arrangements of five colors may make the table shown and then reason that since beginning with red yields 24 arrangements, and there are five possible colors to begin with, the total number of possibilities is 5 x 24, or 120.

As students search for patterns in their data and the class chart, they may notice that the number of arrangements for any number is equal to the number of colors in that arrangement multiplied by the number of arrangements for the preceding number of colors. For example, the number of arrangements possible using four colors is four times the number of arrangements possible using three colors, and the number using five colors is five times the number using four colors.

Number of Colors (Players)	Number of Arrangements
1	1
2	2
3	6
4	24
5	120

The fact that these multipliers increase by one each time may lead students to discover that the number of arrangements is always the product of all the numbers from 1 through the number of colors. For example, the number of arrangements of four colors is 1 x 2 x 3 x 4, or 24. The possible arrangements of any number of objects *(n)* is written *n!*. The symbol ! means *factorial* and is defined as the product of all the positive integers from 1 to a given number *n*. Therefore, 4! (read *four factorial*) is equal to 1 x 2 x 3 x 4, or 24. Recognition of this pattern will allow students to calculate how many arrangements can be made with any number of colors.

Students may use many different ways to predict the number of possible arrangements for 10 players. Some students may extend the chart, multiplying the number of arrangements by the next number of players (colors) in the list and continuing until they reach the number of arrangements for 10 players. Students may also use their understanding of factorials to calculate the correct answer: 10! = 1 x 2 x 3 x 4 x 5 x 6 x 7 x 8 x 9 x 10 = 3,628,800 arrangements.

In the second activity, students are asked to find the number of possible arrangements of seven players with a limitation: one player must always be either first or seventh in the sequence. There are at least two common approaches. In the first approach, students find the number of possible arrangements of six players: 6! = 1 x 2 x 3 x 4 x 5 x 6 = 720. Then they imagine adding the popular player at the beginning and end of the sequences. There must be 720 possible arrangements with the player at the front of the line and 720 with the player at the end, for a total of 1,440 possible arrangements.

In the second approach, students find the number of possible arrangements for seven players: 7! = 1 x 2 x 3 x 4 x 5 x 6 x 7 = 5,040, and then divide by 7 to find the number of those arrangements that begin with a particular player (5,040 ÷ 7 = 720). There must be a like number of arrangements that end with that player. Again, 720 x 2 = 1,440 possible arrangements, given the stated limitations.

For Their Portfolio poses an interesting challenge for students. Students are asked to find the number of possible arrangements for five players who are posing for a team photo. There is one limitation: two of the players refuse to stand next to each other. Students may approach this problem in several ways. In one approach, students can find out the possible arrangements that have the two players standing next to each other and then subtract that number from the total number of possibilities from Part 1. This would result in the following: 120 (total arrangements for all 5 players) minus 48 (arrangements of the 2 players standing next to each other) equals 72 possible arrangements for the team photo.

Blackline Masters

Snapshot

Part 1

A snapshot freezes a moment in time. A series of snapshots becomes a picture record of an event. Can you use a collection of "snapshots" to find out the contents of a box of Snap Cubes?

- Work in a group. There are 22 Snap Cubes in your Snapshot box.
- Take turns shaking and tilting the box so that a Snap Cube appears in the opening. Record the color of the cube. Think of each sample as a snapshot of the contents of the box.
- Keep taking and recording "snapshots" until you think you have enough data to make a prediction.
- Use your data to predict how many of each color there are in the box.
- Be ready to show your data and explain how you arrived at your prediction.

Part 2

What if... you made a second Snapshot box using the number and colors of cubes to match your prediction? If you repeat the experiment with your new box, will the data you collect be close to that collected from the first sampling?

- Work in the same groups, and set up your own Snapshot box. Fill your box with Snap Cubes to match your prediction. Now, tape the lid shut.
- Repeat the experiment from Part 1. Take the same number of "snapshots" that you did with the first box. Record the data.
- Compare the data to that collected from the first box.
- Decide whether you still think the contents of the boxes are the same. Be ready to explain your reasoning.

Suppose that you are writing an article in your school newspaper and you want to find out how students at your school would answer this question: Which would you rather be, very rich or very smart? You want to find out what percentage of students would choose each alternative. Write a summary describing how you would use sampling to gather the data you need. Include the number of samples you would take, how you would make sure the samples were random, and how you would use the data to draw conclusions about students your age.

Color Draw

Part 1

There are 30 Color Tiles in a bag. The tiles are red, blue, green, and yellow. Can you figure out the number of tiles of each color by checking the color of only one tile at a time?

- Work in a small group. Take turns sampling the tiles in the bag by drawing 1 tile from the bag without looking inside.
- Each time you draw a tile, record the color and then return it to the bag.
- Continue sampling the contents of the bag until you are ready to predict which number of each color in the bag.
- Record your group's predictions, the number of each color picked, and the total number of samples you took.
- Be ready to explain how you made your predictions.

Part 2

What if... your bag contains 20 tiles in one of four possible combinations? Could you use sampling to figure out the contents of your bag?

- Work in small groups. Take turns sampling the contents of your bag. These are the possible combinations of contents in your bag

    10 yellow, 5 green, 3 red, and 2 blue
    5 yellow, 5 green, 5 red, and 5 blue
    8 blue, 6 red, 3 yellow, and 3 blue
    8 blue, 8 yellow, 2 red, and 2 green

- Draw 1 tile from the bag without looking inside. Record the color of the tile and then return it to the bag.
- Continue sampling the contents of the bag until you are ready to predict which combination the bag contains.
- Record your prediction, the number of times each color was drawn, and the total number of samples you took.
- Be ready to explain how you used the results of the sampling to reach a decision.

You are on a committee to find out how many students at your school plan to buy tickets to the school play. The information will help you figure out how many performances to schedule. You don't want to ask all of the students in the school if they plan to come to the play. How could you use sampling to make an accurate prediction? Write a summary explaining what you would do to solve the problem.

Even-Steven

Part 1

> Steven has designed three versions of a two-player game called Even-Steven. He wants each player to have the same chance of winning so that his game is fair. Which, if any, of Steven's game variations are fair?
>
> - Players take turns drawing two Snap Cubes from a paper bag. Player 1 scores one point if the colors are the same; Player 2 scores one point if the colors are different. The player with the most points after 20 draws is the winner.
>
> - Predict which of the following versions are fair:
>   Version 1: The bag holds 1 Snap Cube of one color and 2 Snap Cubes of another color.
>   Version 2: The bag holds 2 Snap Cubes of one color and 2 Snap Cubes of another color.
>   Version 3: The bag holds 1 Snap Cube of one color and 3 Snap Cubes of another color.
>
> - Decide who is Player 1 and who is Player 2. Now play Version 1. Here's how:
>   ◆ Put the Snap Cubes in the bag for Version 1.
>   ◆ Take turns drawing two Snap Cubes from the bag.
>   ◆ Record the score, and then put the cubes back in the bag.
>   ◆ Continue until you have completed 20 trials.
>
> - Now change the contents of the bag. Repeat the activity for Version 2 and Version 3.
>
> - Use your data to decide which variations of the game are fair. Be ready to explain your reasoning.

Part 2

> What if... you wanted to design a variation of Even-Steven that uses three colors of Snap Cubes? Use the same rules for scoring as you did in the first three versions. How many cubes of each color would you put in the bag to make a fair game?
>
> - Working with your partner, decide how many Snap Cubes of three colors to put in the bag. Use up to 10 cubes of each color.
>
> - Test the game by making 20 draws and recording your scores.
>
> - If you believe the game is fair, repeat the game at least two more times to collect more data.
>
> - If the game seems unfair, make adjustments to the contents of the bag and try again.
>
> - Be ready to justify why you believe your version is a fair one.

Steven decides to try another version using 4 red and 2 blue Snap Cubes. Without actually playing the game, describe what you expect the results would be. Do you think the version would be a fair one? Write a summary explaining your reasoning.

True Blue

Part 1

Trina wants to win a goldfish at the carnival. In order for her to win, she needs to pick 2 blue tiles out of the "True Blue prize bag," without looking. If the prize bag contains 3 blue tiles and 3 red tiles, what is the probability of winning the game?

- Working in pairs, put 3 blue and 3 red Color Tiles in a bag. Take turns drawing out 2 Color Tiles. If you draw a blue pair, record a "win." Then replace the tiles.

- Predict the number of wins your team will get if you play the game 40 times. Be ready to explain how you arrived at your prediction.

- Now, conduct 40 trials and record the outcomes. Using your results, write the experimental probability of winning *True Blue*. (The experimental probability is the ratio of wins to total trials).

- Compare the experimental probability to your prediction.

- Find the theoretical probability of winning *True Blue*. Here's how:

    (1) Imagine picking the tiles one at a time. To find the theoretical probability that the first tile will be blue, write the ratio of the number of blue tiles to the total number of tiles in the bag.

    (2) Imagine the first blue tile has been drawn. To find the theoretical probability that the second tile will be blue, write the ratio of the number of blue tiles remaining in the bag to the total number of tiles remaining.

    (3) Multiply the theoretical probabilities to find the combined theoretical probability.

- Compare the experimental probability and the combined theoretical probability of winning *True Blue*. Be ready to explain any differences.

Part 2

What if... you wanted to change the game of True Blue so that Trina would have a 1 in 3 chance of winning the goldfish? How would you do this?

- Working with your partner, use up to 10 red and 10 blue tiles. Decide how many of each color to put in the bag. Then calculate the theoretical probability of drawing 2 blue tiles.

- Adjust the numbers of red and blue tiles until the theoretical probability of drawing two blue tiles is 1 in 3. Be ready to explain how you decided how many of each color to use.

- Find the experimental probability of winning your game by conducting at least 30 trials.

- Compare the theoretical and experimental probabilities. Be ready to explain any differences.

Suppose you play *True Blue* with 3 blue, 3 red, and 3 yellow tiles in the bag. Write a brief summary describing how to calculate the theoretical probability of drawing 2 blue tiles. If you were to conduct an experiment of 40 trials, what results would you expect? Explain your reasoning.

Cube Cover-up

Part 1

How many rolls of one pair of dice will you need to place at least one Snap Cube in each of the 36 spaces on the Cube Cover-up game board?

- Using the two colors of your dice, assign one color to represent the top row of dice on the Cube Cover-up game board and use the other color to represent the side row of dice on the board.

- Working in pairs, Player A rolls the dice, and Player B records the roll by placing a Snap Cube on the corresponding space on the Cube Cover-up game board. If the same dice combination is rolled more than once, Player B should stack cubes on top of one another on that space.

- Predict how many dice rolls you will need to cover each space on the board with at least one cube. Record your prediction before you begin rolling.

- When every space has at least one cube, count the total cubes to find out how many times you rolled the dice. Compare your results to your prediction.

- Create a table to show how many spaces are covered by 1 cube, 2 cubes, 3 cubes, and so on up to 6 or more cubes. And, include a column in your table to show the total cubes needed to cover the board.

- Switch roles and repeat the experiment.

- Be ready to discuss your results, predictions, and tables.

Part 2

What if... you picked only 12 spaces on the Cube Cover-up game board that you wanted to cover? About how many rolls of the dice would you need to cover those spaces?

- Working with your partner, circle 12 spaces on the Cube Cover-up game board.

- Predict how many dice rolls you would need to cover your circled spaces.

- One partner rolls the dice and the other partner records using Snap Cubes.

- As soon as you have covered your chosen spaces, count the Snap Cubes on the board to find the total number of rolls.

- Compare the results with your prediction.

- Repeat the experiment a second time. You may use the same 12 numbers or choose different ones.

- Be ready to discuss your results.

For Your Portfolio

In a lottery, players predict which 5 numbers out of the numbers from 1 to 36 will be drawn out at random. Two friends want to enter the lottery. One friend picks the numbers 6-14-23-29-34. The other friend picks the numbers 1-2-3-4-5. Which friend do you think has a better chance of winning the lottery? Explain your answer.

Give and Take

Part 1

Give and Take is a game for 3 players. The object is to be the first player to collect 10 Snap Cubes of a chosen color. Without looking in the bag, can you tell whether each player has an equal chance of winning?

- Work in groups of three. Each player should choose the color he or she wants to collect from the choices listed on the bag.
- Without looking, take turns drawing a cube from the bag. Give the cube to the player who is collecting that color. The first player to collect 10 cubes is the winner.
- When the game ends, record the color that won and the number of cubes collected for each color.
- Play several times, switching colors each time. Continue to record data.
- Without looking into the bag, decide whether the game is fair. Be ready to explain your reasoning.

Part 2

What if... you wanted to change the game of Give and Take to make it fair? You still do not know the exact contents of the bag. What would you change about the game? How can you be sure your version of the game is fair?

- Work in the same groups of three. Using the data collected in the first activity, decide how many cubes would change the contents of the bag enough to make Give and Take a fair game.
- You may add or subtract cubes, but you may not look in the bag. Keep a record of the changes you make.
- Now play the game as you did before. Record which color wins and the number of cubes collected for each color.
- Play the new game several times. Continue making changes, playing the game, and collecting data until you are sure the game is fair.
- Be ready to explain what changes you made and how you know your game is fair.

Suppose you put the cubes back into the bag after each turn. How would the game change? Write a brief summary describing the changes in your results. Be specific about how you would decide if the game was fair.

Collectible Cubes

Part 1

The Jock 'n' Rock trading card company puts a special hologram card in each package of trading cards. There are 6 different hologram cards. The company sends an equal number of each kind of hologram card to every store. How many packages of trading cards do you think you would need to buy to have a good chance of getting all 6 hologram cards?

- Working with a partner, predict how many packages you think you would need to buy in order to get all 6 hologram cards.
- Perform a simulation. Here's how:
  ◆ Put 6 different-colored Snap Cubes in a paper bag. Without looking, pick a cube from the bag and record the color. Return the cube to the bag and shake the bag to mix up the Snap Cubes.
  ◆ Continue picking cubes, recording the colors and returning them to the bag until you have picked all 6 colors. Record the number of picks you made.
- Run the simulation at least 3 times.
- Decide how many packages of trading cards you would need to buy to have a good chance of getting all 6 hologram cards.
- Be ready to explain why you chose that number of packages.

Part 2

What if... the Jock 'n' Rock trading card company decides to put one or the other of two special edition cards in each of its trading card packages? Suppose the company prints 3 times more of the "Jock of the Month" card than the "Rocker of the Month" card. How many packages of trading cards do you think you would need to buy to have a good chance of getting both special edition cards?

- With your partner, make a prediction based on the changes in the situation.
- Design a simulation using Snap Cubes and a paper bag to model the events in this situation.
- Run your simulation at least 3 times.
- Use data from your trials to decide how many packages of cards you would buy.
- Be ready to explain your predictions and simulations, and why you chose that number of packages.

For Your Portfolio

The trading card company wants to increase its number of hologram cards from 6 to 25. A package of trading cards sells for $1.75. Using this information, write a letter to the company president encouraging or discouraging the company from making this decision and give reasons to support your request.

Dizzy Darts

Part 1

If you drop "darts" onto a Tangram dart board, how can you find the probability of landing on each of the colors on the board?

- Working in pairs, make a polygon using 1 set of Tangram pieces. Trace the polygon onto paper. Include the outline of each Tangram piece. This is your Tangram dart board.
- Color the Tangram dart board using as few colors as possible. Make sure that the sections that share sides are different colors.
- Find the *theoretical* probability of a dart's landing on each color. Here's how:
  - Using the small Tangram triangle as a unit of measure, find the area covered by each color and the area of the whole dart board. Then write a ratio like this for each color:

    $$\frac{\text{area of color section(s)}}{\text{area of dart board}}$$

- Now find the *experimental* probability of a dart's landing on each color. Here's how:
  - Drop a dart onto the Tangram dart board and record the color it lands on.
  - Perform 20 trials. Do not count trials that miss the dart board. If your dart lands on the line between two sections, record the color at the centermost point of your dart.
  - Write a ratio like this for each color to show the experimental probability of hitting each section.

    $$\frac{\text{number of times the dart hits the color}}{20 \text{ (total number of trials)}}$$

- Now convert each probability ratio into a percent.
- Make a table to summarize the results of your work. Be ready to explain any differences that you found between the theoretical and experimental results.

Part 2

What if... you wanted to assign points to the different colors on your dart board? How would you do this?

- Use two complete sets of Tangram pieces to design a new dart board. Trace your Tangram shape onto paper, and remember to include an outline of each Tangram piece.
- Color your Tangram dart board using 4 colors. Make sure that pieces that share sides are different colors.
- Compare the areas of the colors. Then assign each color a point value from 1 to 16 points in a way that you think is fair.
- Now find the theoretical probability of landing on each color. Use these probabilities to decide whether you have assigned the points fairly. Make adjustments if necessary.
- Make a chart showing each color, its point value, and the theoretical probability of landing on the color with a dart.
- Be ready to explain the relationship between the points you have assigned to each color and the theoretical probability of landing on that color.

Suppose the theoretical probability of hitting the color red is 1/2. During an actual game, a player hits red 7 out of 20 times. Write a paragraph comparing the theoretical and experimental probabilities. If you were to conduct another experiment of 20 trials, what results would you expect? Explain your reasoning.

Grab Bag

Part 1

Hector insisted that anytime he picked rods out of the class storage box, there would always be a blue rod in his pick. He therefore thought that blue must be the average length for a rod. What do you think? Suppose you reach into a bag and grab a handful of Cuisenaire Rods. What is the typical length of a rod in your handful?

- Each color of Cuisenaire Rod is a different length. The white rod is the shortest at 1 centimeter long; the orange rod is the longest at 10 centimeters long. Note: You may find it helpful to make a chart of rod colors and their corresponding lengths.
- Work with a partner. Put one set of Cuisenaire Rods in a paper bag.
- Player 1 should grab a handful of rods and place them on the desktop, recording the number of each color of rod in the handful.
- Find and record the median length of the handful. Here's how:
    - Organize the rods from longest to shortest. Find the middle rod. If you have an even number of rods, there will be two rods in the middle. If the middle rods are different colors, the median is halfway between their lengths. If the two rods are the same color, the length of that color is the median.
- Find and record the mode length of the handful. Here's how:
    - Sort the handful into piles by color. The pile with the most rods is the color that represents the mode. You can have more than one mode.
- Find and record the mean length of the handful. Here's how:
    - Find the sum of the lengths in the handful. Divide the sum by the number of rods in the handful. Round to the nearest tenth.
- Put the rods back into the bag and have Player 2 grab a handful. Find the mean, median, and mode of the lengths in the handful.
- Now combine the data from both handfuls of rods. Add the rods you grabbed to those in your partner's handful and find the combined mean, median, and mode.
- Be ready to explain what the measures tell you about the typical lengths of rods in a handful.

Part 2

What if... the contents of the bag were changed? Suppose two orange, two blue, and two brown Cuisenaire Rods were removed from the bag. How would this affect the typical length of a rod in a handful of Cuisenaire Rods?

- First predict the results of your experiment. Predict the median, mode, and mean of a handful of Cuisenaire Rods. Be ready to explain your predictions.
- Take turns grabbing a handful of Cuisenaire Rods from the bag.
- For each handful, find the median, mode, and mean of the lengths of the rods.
- Now combine the data from both handfuls and find the median, mode, and mean of the lengths of the combined rods.
- Be ready to explain the results of your experiment and prediction.

For Your Portfolio

The sports writer of a newspaper wrote that a baseball pitcher typically throws 92 miles per hour. What does this statistic really mean? What information would help you to understand the statistic? Write a letter to the sports writer explaining what kind of information readers need to know in order to understand this statistic.

Geo-Hoops

Part 1

Emily invented a game called Geo-Hoops. Geo-Hoops is played much like "Horseshoes," except that instead of ringing a horseshoe around a stake, players toss hoops onto Geoboard pegs. After several rounds, players calculate the mean, median and mode, and the winner is the player with the higher score in at least two of the three statistical measures. How do you think the scoring method will affect the outcome of the game?

- Working with your partner, make five hoops out of pipe cleaners. Here's how:

  (1) Overlap the ends.   (2) Twist them.   (3) Wrap the ends around the hoop.

  Each hoop should be about 3 inches in diameter.

- Place a Geoboard on a flat surface. Mark a throwing line 8 inches away from the board. When it is your turn to toss the hoops, your throwing hand must not cross the line.

- Decide who will go first. Each player gets to toss 5 hoops on a turn. Toss the hoops one at a time, and leave the hoops on the board after each toss. You score points for every Geoboard peg you ring. The pegs are scored as follows:

- Hoops that lean on a peg but do not surround it do not count. If more than one hoop surrounds a peg, count full score for the bottom hoop and half score for the upper hoop(s).

- Each player gets 10 turns. Record the scores for each turn.

- When both players have finished 10 turns, calculate the mean, median, and mode of the scores. (If the middle two scores are different numbers, the median is the number halfway between those scores. If there is more than one mode for a set of data, count the higher one.)

- The player who has the highest 2 out of 3 measures (mean, median, and mode) wins.

- If a tie makes it impossible to declare a winner, both players should take another turn. Calculate the new mean, median, and mode. Continue until a player wins.

- Play two games of Geo-Hoops.

- Be ready to discuss your results.

Part 2

What if... you wanted to find the typical number of points scored on a turn in a game of Geo-Hoops? How could you use the data you have collected to find out?

- Combine the data from both players for each of the games you played.
- Calculate the mean, median, and mode of the combined data.
- Be ready to explain how many points are scored on a typical turn.

The Math Wiz game company wants to sell the Geo-Hoops game in stores. The marketing department needs feedback from students who have already played the game. Write a letter to the company explaining what you think students can learn from playing Geo-Hoops.

How High? How Long?

Part 1

How many white Cuisenaire Rods do you think you can build onto a tower before the tower falls?

- Work with a partner. Find a level surface to use as a base for your tower.

- Take turns stacking white Cuisenaire Rods. Place only one rod at a time. Do not hold the tower while putting a new rod on top. You can straighten the tower between turns. Keep building until the tower falls.

- Record the number of Cuisenaire Rods in the tower. Do not count the final rod in the stack unless it remains in place for at least three seconds.

- Repeat the experiment 20 times.

- Create a frequency graph to display your data.

- Be ready to explain how your data can be used to find out how many white Cuisenaire Rods you and your partner can typically stack.

Part 2

What if... you wanted to find out how far, with one breath, you can blow a white Cuisenaire Rod along a smooth surface?

- Work with a partner. Find a smooth, flat surface.

- Mark a starting place for the white Cuisenaire Rod. Decide how you will measure the distance it will travel.

- Blow the white Cuisenaire Rod along the smooth surface. Experiment a few times to figure out the best angle to blow from.

- Now, work with your partner to complete 20 trials, 10 trials each.

- Record your data on a frequency graph as before.

- Using your data, what distance do you think a "typical" student could move a white Cuisenaire Rod with one breath? Be ready to explain your thinking.

Create your own experiment. Decide which ability you would like to test using Cuisenaire Rods. Work with your partner to complete however many trials you think would be sufficient. Record your data and compare it to the other trials. Write a brief summary of your results.

Rocket Launch

Part 1

Keia just got a new pool table. She noticed that when one ball hits another ball most of the momentum of the first ball is transferred to the second ball, causing it to move. She wants to see if the same idea will work with Tangrams. Make a Tangram rocket and launcher to simulate Keia's pool table. Using momentum, how far can you make the rocket travel?

- Work with a partner. One person will launch the rocket; the other person will measure the distance it travels.
- Find a smooth surface. Make a small mark on the surface and position the edge of a Tangram square next to the mark. Place the rocket (small Tangram triangle) and the launcher (medium Tangram triangle) as shown at right.

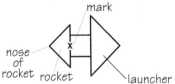

- To launch the rocket, hold the square firmly in place; then slide the launcher several inches away from the square and back again so that you strike the square sharply. The rocket will move away from the square as the momentum transfers.
- Measure the distance from the small mark you made on the surface to the nose of your rocket. Record the measurement. Round your measurement down to the nearest whole inch.
- Repeat the experiment until each player has launched the rocket 20 times.
- Create a frequency graph to display your combined data from 40 launches.
- Be ready to explain the typical distance a small triangle Tangram rocket travels.

Part 2

What if... the rocket were a medium Tangram triangle and the launcher were a large triangle? What do you predict would be the typical distance your new rocket would travel?

- Working with your partner, compare the sizes of the rocket and the launcher in the first activity. Then compare the sizes of the new rocket and its launcher.
- Each person should conduct one sample launch to "feel" how the change in pieces may affect the results of a launch.
- Discuss your findings and make a prediction. Be ready to explain how you arrived at your prediction.
- Conduct 40 launches, 20 by each partner. Measure the distance traveled as before, rounding down to the nearest whole inch.
- Create a frequency graph of your data. Using your data, find the typical distance the rocket traveled.
- Compare your results to your prediction. Be ready to explain any differences.

Two students perform 40 launches using a small Tangram triangle for the rocket and a large Tangram triangle for the launcher. The greatest distance traveled was 44 inches; the smallest distance was 6 inches. To find the typical distance their rocket traveled, they averaged 44 and 6 by adding them and then dividing by 2. Do you think this method is a good one for finding the typical value in a set of data? Write a brief summary explaining your reasoning. Use examples from your experiments to support your answer.

Freeze Before Fifty

Part 1

Freeze Before Fifty is a game for 2 players. The object is to collect as close to 50 Snap Cubes as you can without going over. Can you find the winning strategies?

- Work with a partner. Place 100 Snap Cubes where both players can reach them.
- Both players roll the 2-2-2-3-3-4 number cube once and collect that number of Snap Cubes. These are your "starting numbers."
- Decide who will go first. On your turn, spin the Freeze Before Fifty spinner and roll the number cube to get an operation and a number. If you spin addition, add the number you roll to the number in your collection. (On your first turn, this number would be your starting number). If you spin multiplication, multiply the number you roll by the number of cubes you already have. Collect that number of cubes and add them to your pile.
- Take turns spinning and rolling to increase your collections.
- If you reach more than 50 Snap Cubes on any turn, return all your cubes to the pile and roll a new starting number. On your next turn, start building a new collection of cubes.
- When you think you are as close to 50 as you can get without going over, say "freeze" when your turn comes. Do not add more cubes. The other player gets up to two turns to get closer than you are to 50. The other player can stop at any time.
- Once one player freezes and the other completes his or her final turns, the game is over. The player with the number of cubes closer to 50 without going over wins. If there is a tie, there is no winner, and players should start the game over.
- Play several games of Freeze Before Fifty; take turns going first. Be ready to discuss your strategies for winning.

Part 2

What if... you change the range of numbers that you could roll? How would this change your game strategies?

- Start the game in the same way. Roll the number cube to choose a starting number.
- Then, roll both dice, the number cube (marked 2 to 4) and the standard die (marked 1 to 6), and find the sum of your rolls.
- Now spin the spinner and add or multiply your total roll by the number of cubes in your collection. Collect that number of Snap Cubes and add them to your pile.
- Continue playing (as you did in Part 1) until there is a winner.
- Play several games of this version of Freeze Before Fifty.
- Be ready to explain how changing the range of numbers affected your game strategies.

Suppose you are playing Freeze Before Fifty with a die marked with the numbers 1, 1, 1, 2, 3, and 4. How do you think this change would affect the play of the game? How might this change affect your strategies? Write a brief summary explaining your reasoning.

Half Chance

Part 1

Half Chance is a game for 2 players. The object is to be the first player to cover half of the game board with Cuisenaire Rods. Can you find the winning strategies?

- Choose which side of the Half Chance game board you will play and then spin Spinner A. The player who spins the higher number goes first.

- On your turn, you can either spin to add Cuisenaire Rods to the board or spin to remove them. You must decide before you spin.

- *Adding Cuisenaire Rods:* Decide whether to use Spinner A, Spinner B, or a combination of the 2 spinners and then spin twice. The outcome of the first spin tells which color Cuisenaire Rod to add to the board. The outcome of the second spin tells how many of that color you may add to the board. For example, if you spin light green on Spinner A and 2 on Spinner B, you add 2 light green rods.

- *Fraction Bonus Turn:* If at the end of your turn, you have covered exactly 1/4 (25 squares), 1/2 (50 squares), or 3/4 (75 squares) of your portion of the game board, you take an additional turn.

- *Placement of Cuisenaire Rods:* Once a rod is played, it can't be moved unless it is removed from the board during a player's turn. Added rods may be played in any empty space. If you can't add all the rods indicated by your spin, you add no new rods and instead remove one rod of any color from the board.

- *Out of a Color:* If you do not have enough Cuisenaire Rods of the color spun to make your play, spin both spinners again.

- Play continues until 1 player covers his or her half of the game board.

- Play several games of Half Chance. Be ready to discuss winning strategies.

Part 2

What if... you change the game of Half Chance so that added Cuisenaire Rods must connect? How would this change your game strategies?

- Play the game again, except this time any rods that you add must connect along a side to form a block or a chain.

Okay:    Not okay:

- Be ready to explain how the change in the rules affected your game strategies.

Write a letter to a new student who will be playing Half Chance for the first time. Give the student advice on the best strategies to use. Include your strategies for choosing which spinners to use, placing rods, and earning Fraction Bonus turns.

Block Path

Part 1

Block Path is a game in which players build paths of Pattern Blocks. The object of the game is to build a path that ends closest to the finish line without going over it. Can you find the winning strategies?

- Work with a partner. Each of you needs a paper track that is about 70 centimeters in length. Mark a starting line at one end of the paper. Mark a finishing line 50 centimeters away from start. All moves must fit inside this strip of paper.
- Roll a die and collect that number of one kind of Pattern Block shape.
- To begin your path, place your first block at the starting line. Then continue to place blocks so that each new block touches the last block you placed. Each block must be closer to the finish line than the previous block.

- After your partner has taken a turn and added to his or her own path, roll again and take that number of a different Pattern Block shape. Add them to your path without rearranging the blocks that you placed on your previous turn(s).
- Continue playing until each of you has had six turns, and you have used each kind of Pattern Block shape.
- The player whose last piece is closest to the finish line without going over it is the winner. If both of you go over the finish line then the player whose last piece is closest to the finish line is the winner.
- For each player, record the total distance (in centimeters) from the starting point to the end of the path and the total number of each shape used by each player.
- Play several games of Block Path.
- Be ready to talk about your strategies.

Part 2

What if... you use two dice instead of one, spell out the word for the number rolled, and use that many Pattern Blocks to make your path? How will this change your game strategies?

- Start playing Block Path again, but this time roll a pair of dice. For each roll, add the numbers on the dice and then figure out the number of letters in the number you rolled. For example, for a sum of ten, you collect 3 Pattern Blocks because there are 3 letters in the word ten. For a sum of eleven, you collect 6.
- Collect and place blocks into a pathway as you did in Part 1.
- The winner is determined in the same way as before. The game is over after each of you has had 6 turns.
- As before, record the total length of the paths in centimeters and the total number of each shape used by each player.
- Play several games of this version of Block Path.
- Be ready to discuss how the change in the rules affected your game strategy.

Games that are played with dice usually involve chance, but many, such as Block Path, also have an element of skill. Think of a game that you have played that requires dice. Write a summary explaining whether the game is more a game of chance or a game of skill. Use specific examples from the game to prove your point.

Mall Madness

Part 1

> Dominic is an architect who plans to have each level of a shopping mall painted red, blue, green, or yellow. The mall has three levels. Dominic wants each level to be a different color. How many possible color combinations are there for the three levels? How many different ways can the colors be arranged?
>
> - Working with a partner, use Color Tiles to represent the different levels of the shopping mall. Find all the possible color combinations for the three levels. Order doesn't matter in a combination. *Red-blue-green* is the same as *green-blue-red*.
>
> - Make a list of the combinations. Make sure your list does not contain repeats of the same combinations.
>
> - Now find all the ways the colors can be arranged. These are called permutations. Order does matter in a permutation. *Red-blue-green* is therefore not the same as *green-blue-red*.
>
> - Record each permutation. Find a way to organize your solutions.
>
> - Compare the numbers of combinations and permutations. Be ready to discuss their relationship.

Part 2

> What if... the levels of the mall do not have to be different colors? How many color combinations are there? How many ways can the colors be arranged?
>
> - Working with your partner, use Color Tiles to find the possible 1-color, 2-color, and 3-color combinations. *Red-red-green* and *green-green-red* would both be considered a single 2-color combination of *red-green*.
>
> - Find ways to record your solutions.
>
> - Now use the Color Tiles to find all the possible permutations. Find ways to record your work.
>
> - Be ready to explain the relationships between the number of combinations and permutations.

For Your Portfolio

> Five students have worked together to perform a science experiment. The teacher will choose two students from the group to present the experiment to the class. Write a summary explaining the difference between a list of combinations and a list of permutations. Which list do you think would be more helpful to the teacher in this situation? How many possible combinations and permutations would the teacher have to choose from in order to select two students from the group? Include a list and/or any charts or diagrams that might be helpful for the teacher to see every possibility.

Quilt Squares

Part 1

Quilt squares are made from cloth geometric shapes that are pieced together. Once a pattern is chosen, the quilt maker sometimes varies the colors of the shape to create an interesting design. Can you find all the possible arrangements of color that can be used to fill a quilt square?

- Work in a small group. Each of you needs a copy of the *Quilt Squares Patterns* page. The quilt maker plans to use white fabric for the center square. The triangle pieces that border the square can be red, blue, yellow, or green. But there can be no more than two different colors of triangles in any quilt square.

- Use your small Tangram triangles to find all the possible arrangements of two colors. Check to make sure you have not created any repeats. Turning a quilt square does not count as a different quilt square.

  Example: These should be considered one quilt square.

- Record all the possible quilt squares on your *Quilt Squares Record* page by coloring or labeling the triangles.

- Count the total number of different arrangements you have found. Be ready to share all the possible quilt squares that can be made with two of the four possible colors.

Part 2

What if... you can use up to four colors of triangles in one quilt square? How many color arrangements are there? Can you find them all?

- Make a prediction as to how this change in the rules will affect your work. How many different quilt squares do you think there will be using up to four colors of small triangles? Record your prediction.

- Work with your group to find all the possible quilt squares. Use the small Tangram triangles and the *Quilt Squares Patterns* page.

- Remember, you can use up to four colors in each quilt square. Each arrangement of colors must be different. Use the *Quilt Squares Record* page to record your solutions.

- Be ready to explain how you know that you have found all the possible solutions.

 A circular stained-glass window has a border of panes made from 5 colors. How would you go about finding the number of possible arrangements that you can make using all 5 colors. Turning the border does not count as a different arrangement. Write a summary explaining how you would go about finding the number of possible arrangements there would be for 5 colors.

Starting Five

Part 1

The Long Tall basketball team has five star players. At every game, the players want to run out onto the floor in a different order. How many possible orders, or arrangements, of players are there? Can you discover a pattern that will help you figure out how many ways there are to arrange any number of players?

- Working with a partner, use 5 different Snap Cube colors to represent the five basketball players.
- Find all the possible ways to arrange the five players as they run out onto the court one at a time.
- Find ways to record your solutions, first giving the number of possible arrangements for 1 player, 2 players, 3 players, 4 players, and 5 players. Make sure you do not repeat the same arrangement of players.
- Look for a pattern in your data that would help you predict the number of possible arrangements of any number of players. Predict the number of different arrangements there are for 10 players.
- Be ready to discuss the number of arrangements for each number of players.

Part 2

What if... there are seven star players on the team, but one player has become so well-liked by the fans that this player always runs out onto the floor either first or last? How many possible arrangements of the players are there now?

- Use different colors of Snap Cubes to represent the players.
- Find all the possible ways to arrange seven players if one player has to be either first or last every time.
- Find ways to record your work. Make sure you don't repeat the same arrangement of players.
- Is there a way to find the answer without making and recording every combination? Try it. Look for patterns.
- Be ready to explain how you got your answer.

The team decides to have a photo taken of its five starters, but the two guards refuse to stand next to each other. Suppose you need to find the number of possible arrangements of players. How would you do it? Write a brief summary explaining how you would organize your work and what calculations you would perform to find an answer.

©Cuisenaire Company of America, Inc. Probability/Statistics ◆ Grades 7-8

Cube Cover-up Game board

Freeze Before Fifty Spinner

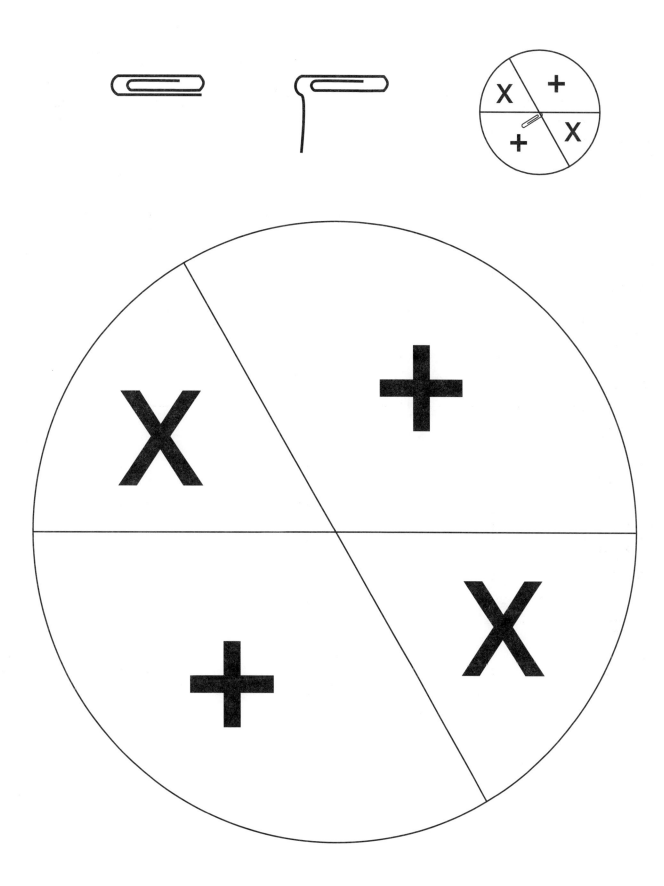

©Cuisenaire Company of America, Inc.   Probability/Statistics ◆ Grades 7-8   **115**

Half Chance Game board

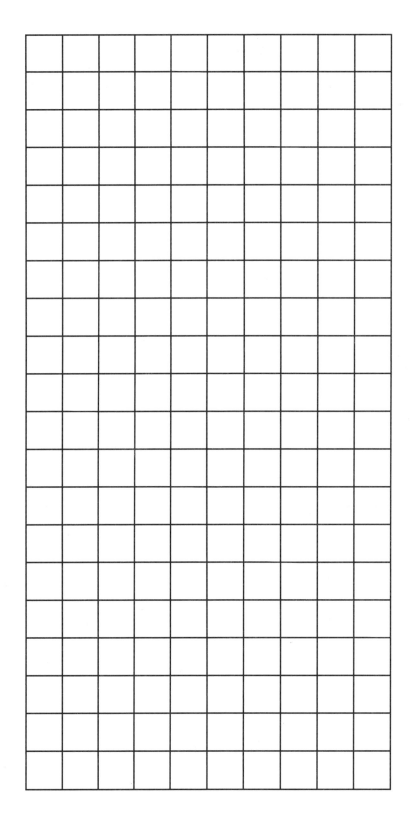

Half Chance Spinner A

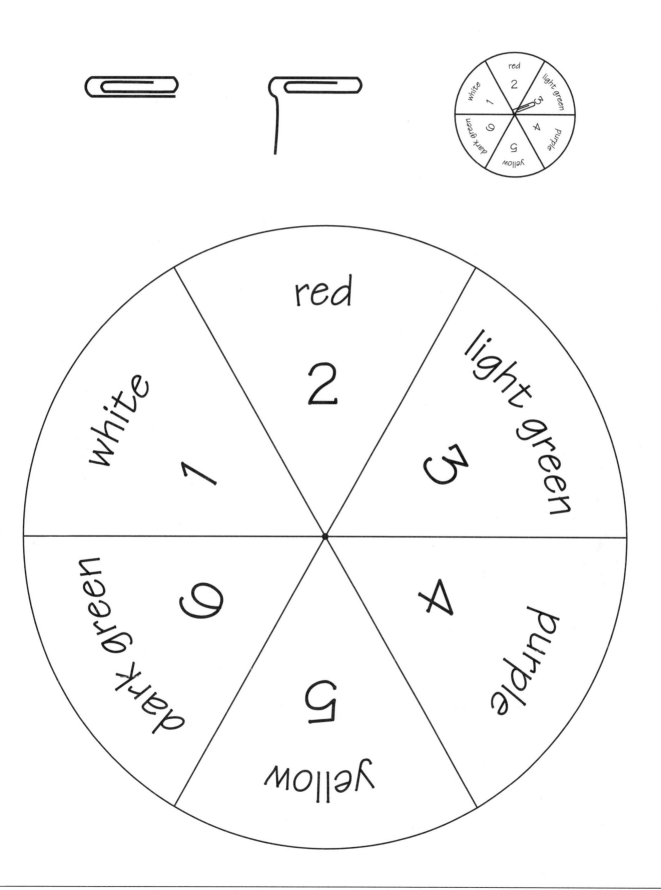

Half Chance Spinner B

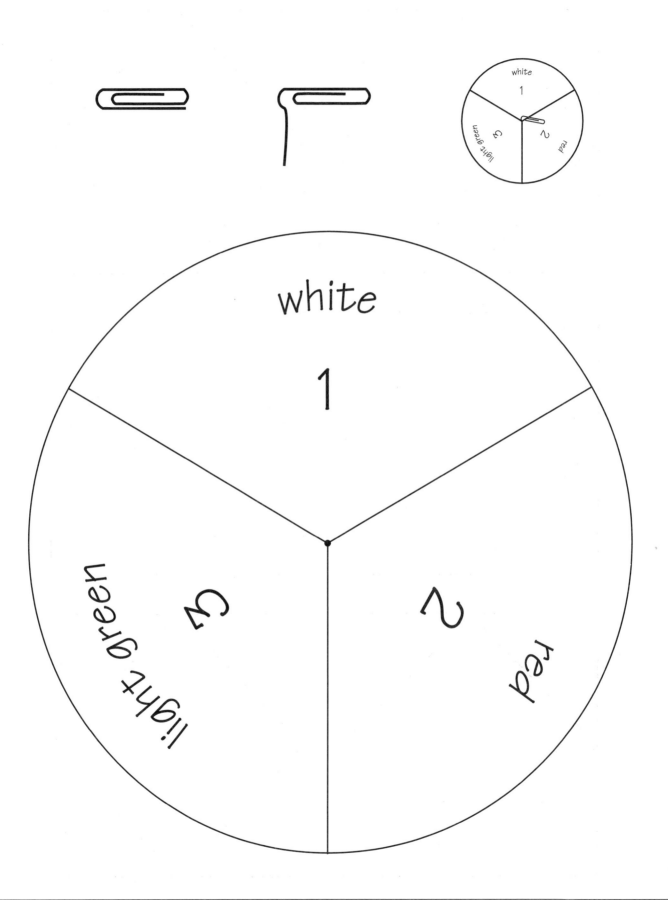

118 the Super Source • Probability/Statistics • Grades 7–8 ©Cuisenaire Company of America, Inc.

Quilt Squares Patterns

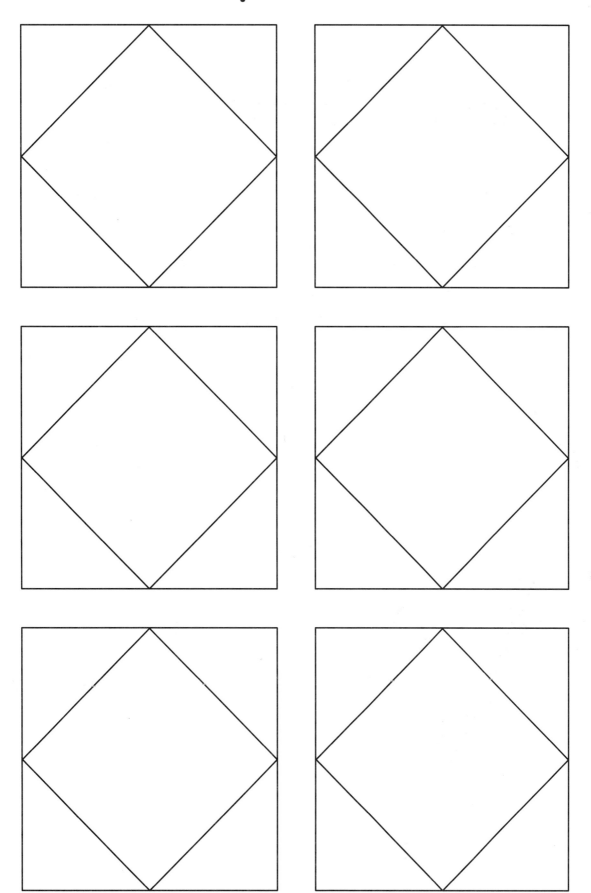

Quilt Squares Record